ISLAND FARMING

History and Landscape of Agriculture
in the San Juan Islands

Boyd C. Pratt

Mulno Cove Publishing
FRIDAY HARBOR

*Between weather & bugs & democracy
the farmer has a rocky time of it.*

— C. E. Cantine

ISLAND FARMING
History and Landscape of Agriculture in the San Juan Islands

by Boyd C. Pratt

Copyright © 2022 Boyd C. Pratt
All rights reserved

Second Edition

No part of this publication may be reproduced or transmitted in English or in other languages, in any form or by any means, electronic or mechanical, including photocopying, digital scanning, recording, or any other informational storage or retrieval system, without the written permission of the author.

ISBN: 978-1-7342351-2-8

Published and Distributed by Mulno Cove Publishing
Printed in the United States of America

Table of Contents

Preface i
Introduction iii
Cultivating the San Juan Islands
    Physical Features 1
    Climate 4
    Coast Salish Farming 7
    Belle Vue Sheep Farm 10
    What about the Pig? 23
    Dividing Up the Land 25
    Homesteading 33
    Open Range 48
    Commercial Farming 54
    Sheep 62
    Hay and Grains 67
    Fruit 73
    Dairy 86
    Poultry 98
    Peas 103
    Miscellaneous Crops 104
    Scientific Farming 110
    Mid- to Late Twentieth Century 116
    Into the Present 121
Touring the Agricultural Landscape of the San Juan Islands
    Introduction 131
    Inter-Island Ferry Route 134
    Lopez Island 137
    Orcas Island 151
    San Juan Island 161
Suggested Reading 187
Photograph and Illustration Credits 193
Acknowledgements 196
Appendices
    Glossary 197
    Plants and Animals 202
    San Juan County Extension Agents 205
    Island Farmers 206

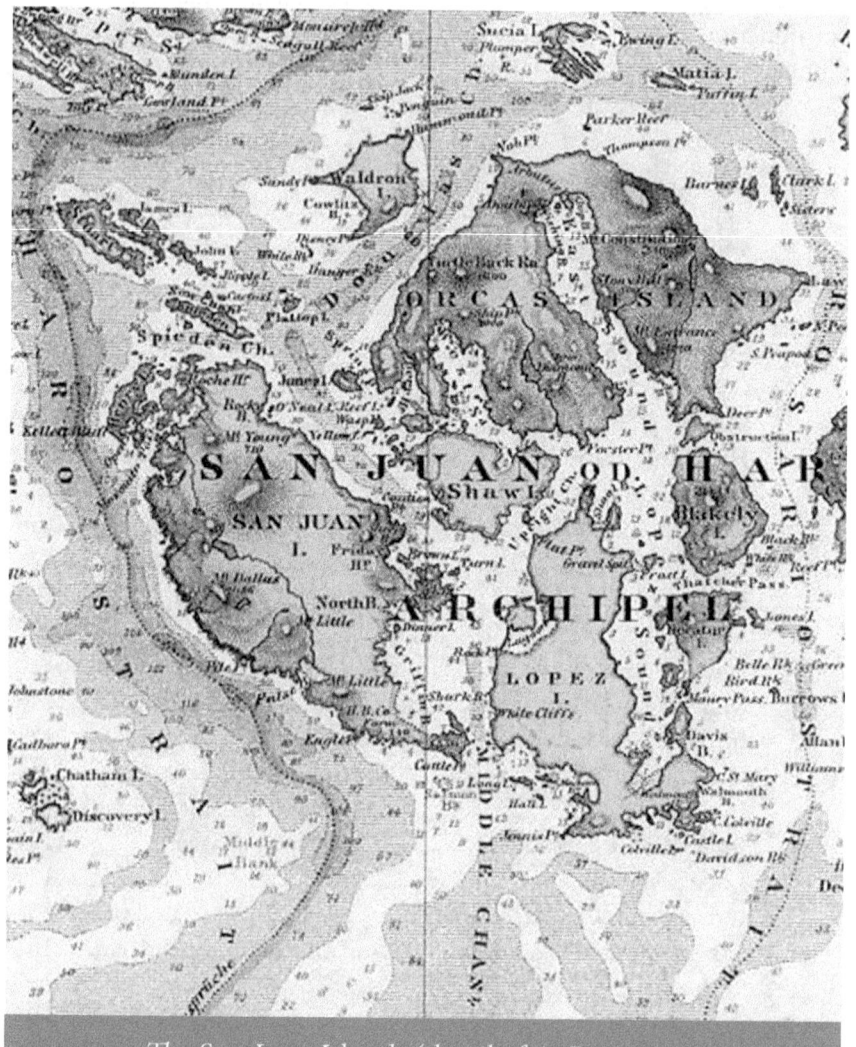

The San Juan Islands (detail of A. Petermann Karte des San Juan OD. *Haro Archipels, 1872)*

PREFACE

Lovel and I started a small, diversified farm on San Juan Island in 1991. The core of our operation was a CSA (community supported agriculture)—the first one in San Juan County. While working on our farm, I became interested in the history of agriculture in the islands. In particular, I wanted to know about the pig that Lyman Cutlar shot on June 15, 1859, the event that precipitated the Pig War. What breed was it? Why was it ranging free, rooting in Cutlar's potato patch? And, for that matter, what kind of potatoes were in the patch, and why wasn't the patch properly fenced? These questions led to an intensive study of the archival material surrounding the pig incident, and eventually the transcription of Chief Trader Charles Griffin's *Belle Vue Sheep Farm Post Journals*, a detailed account of a Hudson's Bay Company farm.

Island farming has been a story of change from the very beginning. The Coast Salish manipulated the environment to foster camas growth and sequester their prized woolly dog stock. First Hudson's Bay Company, then American homesteaders, transformed these agricultural environments for their own purposes. The economics of farming—always marginal—posed challenges for island families, but they used the advantage of water transportation to market their crops to surrounding urban settlements. Many times when Lovel and I were farming did we hear the refrain, "You can't make a living farming in the islands today." Historically, it was also extremely rare that raising crops was the sole source of income for farmers: homesteaders cleared their land and sold the wood for steamers and the lime kilns; pioneers got jobs as postmasters and road supervisors; and farmers even bought boats and freighted goods and passengers around the Salish Sea.

While one could view the period from 1890 to World War II as the golden age of agriculture in the islands—when it seemed that a farmer could make a decent living cultivating one or two cash crops and large farms such as Bellevue, GEM, and the Orcas Fruit Company shipped thousands of dollars' worth of produce off of the islands to be sold (some even existing comfortably off of one acre's production of rhubarb!)—this view is based on exceptional cases, not the norm. Farmers are a breed that look toward the advantage and adapt accordingly—or else they don't survive. And on islands, all the more so.

Recently Lovel and I were driving on the south end of Lopez Island, field-checking the Tour Guides portion of this book. We rounded one of the corners of the small "doglegs," or right-angled turns, and suddenly found the road in front of us filled with sheep. They were being herded by a sheep dog and carefully watched over by a guard llama. Around the next corner came a truck with the shepherd—an old farming friend of ours. We braked aside each other, windows open, and he said, "Boyd, what are you doing here?"—as if the oddity of the situation was my presence, not the sheep blocking a county road. As the flock parted and passed around us, buffing up the car as they trotted by, it came home to us that the history of agriculture in the islands was still quite present.

The purpose of this book is to celebrate those brave and enterprising men and women who made island farming what it was and is and will be.

PREFACE TO SECOND EDITION

I thank all my readers. Many of you have provided additions, corrections, and comments that have been incorporated into this second edition, along with an additional tour site, a short section on mink farming, and an index—all toward cultivating a more accurate and accessible book.

INTRODUCTION

> Almost by definition an inhabited landscape is the product of incessant adaptation and conflict: adaptation to what is often a new and bewildering natural environment, conflict between groups of people with very dissimilar views as to how to make that adaptation. The political landscape, artificial though it may be, is the realization of an archetype, or a coherent design inspired by philosophy or religion, and it has a distinct purpose in view. But the inhabited landscape is, to use a much distorted word, an existential landscape: it achieves its identity only in the course of existence. Only when it ceases to evolve can we say what it is.
>
> — John Brinckerhoff Jackson,
> "A Pair of Ideal Landscapes"

This book explores the history of agriculture in the San Juan Islands, from the time of the Coast Salish habitation (ca. 5,000 BCE) to the present, assuming a definition of agriculture as the cultivation of animals and plants for food and other products such as fiber and medicine. Seafood and other marine crops are not included in this history; the distinction between agriculture (cultivation on land) and mariculture (cultivation of animals and plants from the sea) has been observed and maintained, in part because the history of fishing in the islands is the subject of another book. Although the broader San Juan Archipelago extends north into Canada's Gulf Islands and east into Washington State's Whatcom County, with many shared physical and cultural properties, in this book's history the San Juan Islands refers to those islands within San Juan County in Washington.

Island farming is both the same and different from mainland farming. On islands, farmers grow the same crops and use similar agricultural techniques as on the mainland; after all, those who are not native to the region bring their cultural—and thus agricultural—traditions along with them. Beyond subsistence, island farmers generally grow for an off-island market. But they also must adapt to the unique physical environment of each island as well as deal with the difficulties surrounding transporting goods off-island to market. Islands offer farmers both challenges and opportunities unique to their environs.

To understand the physical environment in which various cultural groups farmed, this book begins with an examination of the geology and geography of the San Juan Islands: how the underlying strata of the islands was formed over millennia and then refashioned by serial glaciations. This geography not only determined the soil types and subsequent vegetation of the various island ecosystems, but also the climate: the rain shadow—a drier area in the lea of a mountainous region—produced by the surrounding Cascade and Olympic Mountain ranges determines the islands' annual average rainfall as well as the pattern of precipitation and winds. The San Juans have a temperate climate, in part because they are surrounded by water, which offers a moderating influence on ambient temperatures.

The islands were first permanently inhabited by the Coast Salish peoples around five thousand years ago—about the time of the northward advance of the forests after the retreat of glaciers from the region. They practiced several forms of agriculture—camas cultivation and harvesting and woolly dog breeding and shearing—within their traditional use areas, which included the islands. After the Hudson's Bay Company arrived in the area in the early 1840s and established Belle Vue Sheep Farm on San Juan Island in 1853, EuroAmerican farming techniques were introduced—taking advantage of the pastureland on savannahlike prairies that had been maintained through Coast Salish flash-burning. It was the shooting of a Hudson's Bay Company pig by an American settler that precipitated the Pig War (1859), a dispute over differing interpretations of the boundary between England and the United States that ignored Coast Salish territorial claims, and the subsequent Joint Occupation (1860–1872) of the Disputed Islands.

From this early, EuroAmerican agricultural use of the land emerged the homesteading period in the islands' history, beginning with the settlement of the boundary dispute (in favor the United States' claim) in 1872 and the imposition of an American township-and-range land division system in the mid-1870s. Americans—many of them former British subjects who naturalized—established small-scale, self-sufficient, diversified farms on 160-acre homesteads that they either purchased or earned through "proving up" from the US government.

Within a few decades, these farmers began specializing in two or three crops grown for market. In the days of relatively easy marine transportation—with so many steamers plying the waters of the Puget Sound and Salish Sea, they were called the Mosquito Fleet—products were delivered to regional port cities such as Bellingham, Olympia, Seattle, Tacoma, Vancouver, and Victoria. At this point, the narrative of island farming focuses on the commercial farming of several of these major crops: sheep; hay and grains; fruit; dairy; poultry; and peas; along with some unexpected and intriguing specialty crops such as ginseng, rhubarb, and tulips.

This mode of commercial farming lasted well into the twentieth century, changing with the new agricultural techniques advocated by an approach based on science and affected by the regional economic restructuring produced by the New Deal and World War II. Some farmers responded to these changes and adapted. After a period of decline in the number of farms and farmers, the amount of land farmed, and crops produced during the mid-twentieth century, agriculture once again changed. In some cases, farms have thrived in response to the modern-day, tourism- and retirement-related economics of the islands. While it is harder for historians to analyze the times they live in than to seek out trends in the past, it is clear from just observing the landscape that island farming, while in many ways quite different from that of the past, is still a vital part of the islands' economy and way of life.

Throughout this book, the emphasis is on the material culture of farming: how farmers cultivated their fields, with what tools, and with what sorts of structures. These methods, implements, and buildings are what make agriculture work. But above all the men and women who did the work—the farmers—are the protagonists of this history. These factors—along with the unique physical conditions of the San Juan Islands—made island farming what it was and is.

This book is divided into two major sections: a history—titled **Cultivating the San Juans**—and tour guides—**Touring the Agricultural Landscape of the San Juan Islands**. It does not have footnotes, but there is a section called **Suggested Reading** that lists most of the relevant sources I used. There are also several appendices: a **Glossary** of agricultural and farm-related terms; a list of **Animals and Plants** (with Latin names) that includes the species referred to in the book; a roster of **San Juan County Extension Agents**; and short biographies of **Island Farmers** mentioned in the book. There are farmers, past and present, who are not named here. My apologies to them and their families for not including everyone who played a significant role in the history of farming in the islands.

Some comments about grammar and spelling are in order. For most of the quotes, the original wording is maintained, in order to retain more feeling for the times and circumstances. Some names which appear to apply to the same place are for different places and therefore apparently inconsistent; for instance, East Sound is a body of water near Orcas Island while Eastsound refers to the town to its north (the settlement of West Sound, however, was never conjoined). Despite common usage (and the urgings of the editor), breeds of animals and varieties of plants are capitalized—who could resist calling out "Mortgage Lifter" as a potato variety? The same can be said of some titles, such as San Juan County Extension Agent or Master, San Juan Island Grange #966. Finally, the date of homesteads and preemptions is that of the award of certification, which occurred years after the original claim and in some cases after the death of the claimant—a source of confusion.

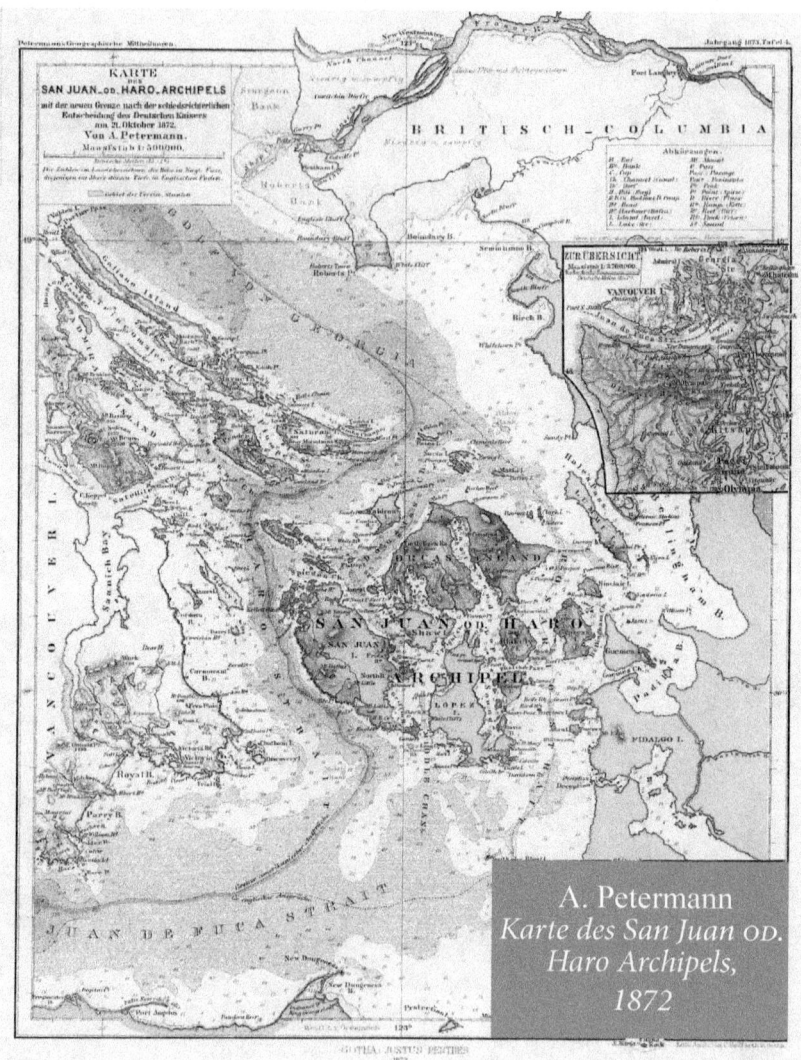

A. Petermann
Karte des San Juan OD. *Haro Archipels,*
1872

CULTIVATING THE SAN JUAN ISLANDS

Physical Features

The physical features of the San Juan Islands—geography and geology, soil conditions and climate—have a signal effect upon local farming. San Juan County consists of over four hundred islands, reefs, and rocks at high tide. Located in the northwest corner of Washington State, these islands are bounded by Haro Strait to the west and northwest, the Strait of Georgia to the north and northeast, Rosario Strait to the east, and the Strait of Juan de Fuca and the Puget Sound to the south. This region lies within the recently named Salish Sea, defined as the body of water surrounded by landmasses of the Olympic Peninsula to the south, Vancouver Island to the west, and the US and Canadian mainland to the north and east. San Juan County—which encompasses most of the San Juan Archipelago—has more than 400 miles of marine shoreline and approximately 172 square miles of land surface, with the three largest islands comprising about 80 percent of the overall land mass: Orcas (57 square miles or 36,432 acres), San Juan (55 square miles or 35,448 acres), and Lopez (29 square miles or 18,847 acres); the next largest islands—Shaw, Blakely, Waldron, Decatur, Stuart, and Henry (in descending order)—each have 10,000 acres of land or less.

The geology of the islands is extremely complex; in essence they comprise the highest points of a submerged mountain range, consisting of older base rock associated with the surrounding Cascade and Olympic Mountain ranges. During the late Pleistocene Era (occurring 50,000–12,000 years ago), a series of three great glaciations occurred, with the ice reaching as far south as modern-day Olympia, filling the depression known as the Puget Trough and leaving only the surrounding mountain ranges uncovered. Upon their recession, the glaciers not only scraped the existing rocky areas, but also left behind glacial till, as well as the large boulders, called erratics, that can be seen standing, unmoved, in the middle of farm fields and pastures.

After the release of weight due to the melting of glacial ice, the islands gradually rose in a process called elastic rebound. During the same period, the melting of the receding glaciers also caused a gradual rise in sea level, so that the result of these forces led to a series of shorelines that differed from the current one. At one point, the shoreline was four hundred feet above what we find today. All or most of Beaverton, San Juan, and West Valleys and other lower

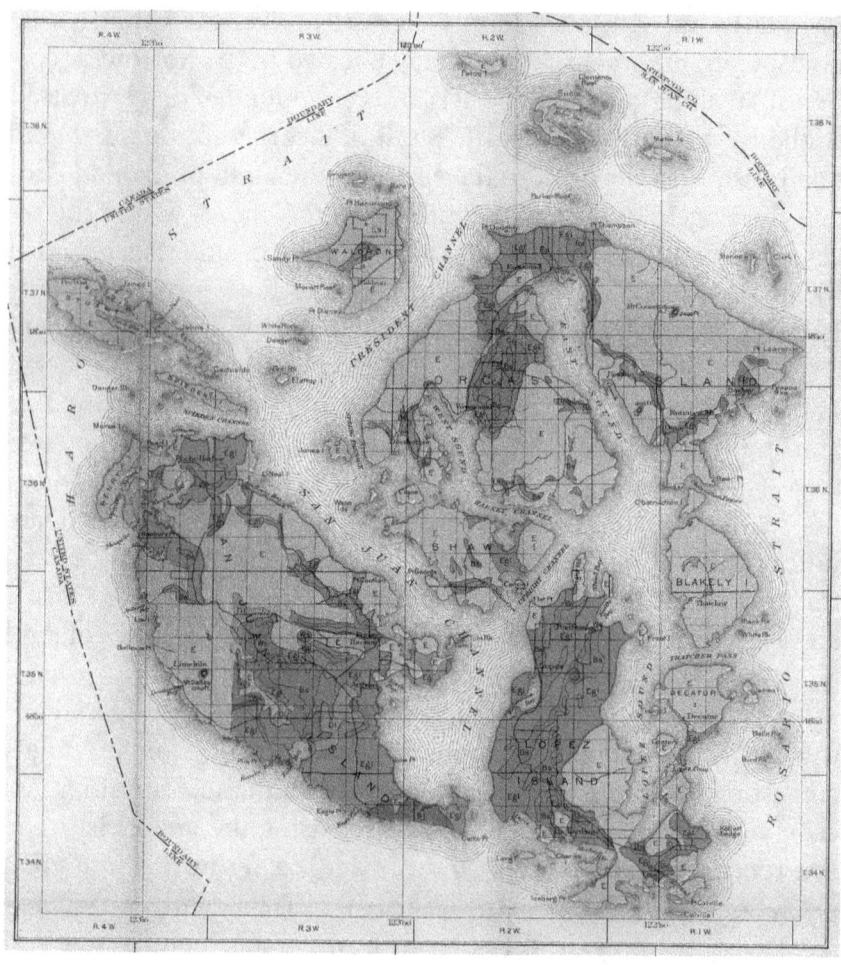

1910 Soil Map of San Juan County

areas on San Juan Island; Deer Harbor, Crow Valley, and Eastsound on Orcas Island; and major portions of Lopez Island were under water at one time. This, together with the glacial action, resulted in the current general soils of the Bellingham-Coveland-Bow Association: low-sloped, poorly drained soils interspersed with small outcrops or "islands" of Roche-Rock Complex (rocky areas that have no soil cover and therefore cannot be farmed). These soils typically have a profile of 18–30 inches above relatively impermeable clay subsoil that is locally referred to as "hardpan." The other soil association—Roche-San Juan—occurs in the uplands and other rocky settings with better drainage. These soil associations led to a pattern of crops in the lower areas and pasture in the upper areas, with farm buildings located on nonarable areas such as rocky outcrops.

The islands exhibit a wide degree of relief. There are numerous rocky knobs, including 15 mountains that rise 1,000 feet or more above sea level, with Mount Constitution on Orcas being the highest at 2,409 feet in elevation. The glaciations of the islands led to areas of low relief, such as glacial plains and gently rolling and basinlike areas, exemplified by San Juan Valley on San Juan, Crow Valley on Orcas, and several parts of Lopez. Drainage occurs by means of short, intermittent streams.

After the glaciations ended some 12,000 years ago, the islands were essentially devoid of plants and animals; the landscape that first developed consisted of open, savannahlike prairies. Western red cedar, Douglas fir, and their respective plant communities arrived as recently as 3,000 to 4,000 years ago. The more extensive prairies contained well-drained soil with scattered Garry, or white, oaks, while the forested communities were largely composed of Douglas fir mixed with western hemlock, white fir, Sitka spruce, and western white pine. Deciduous trees such as red alder and bigleaf maple were also present, particularly in disturbed areas. Lower elevations contained stands of Douglas firs mixed with lodgepole pine. While wetter soils supported growth of cedars, some of the drier areas saw a scattered growth of madrona.

San Juan Valley from Little Mountain, San Juan Island

Climate

The climate in the San Juan Islands is generally mild, tempered by the surrounding seawaters and westerly winds. Due to the rain shadow caused by surrounding mountains on Vancouver Island to the west, the Olympics to the south, and the Cascades to the east, the number of days of sunshine is high compared to the surrounding region. As the term implies, this rain shadow also causes the San Juans to have a lower average rainfall of 29 inches per year (based on the station at Olga on Orcas Island, which has recorded data from 1890 to the present)—compared to the Olympics, which experience over six feet of rain each year. However, rainfall varies across the archipelago, with the general rule being wetter in the north and drier in the south. On San Juan Island, for instance, precipitation ranges from a low of 19 inches per year at Cattle Point to the south to 29 inches per year at Roche Harbor to the north, with San Juan Valley in between averaging 25 inches per year. Elevation, of course, also makes a difference: the lowest-lying lands on Lopez average 19 inches per year, while Mount

Constitution on Orcas has been known to get 45 inches per year. During the period of record, average annual precipitation has varied from almost 38 inches in the wettest year (1917), to 15 inches during the driest year (1929). On average, the 30-year period from 1891–1921 was wetter (31.04 inches per year) than the succeeding 24 years, 1921–1945 (26.33 inches per year). The period following World War II and extending to 1975 witnessed a trend toward wetter and cooler years—factors that farmers had to take into account in their practices.

Average daily temperatures range from 40° F in the winter to 59° F in the summer. The islands have a historic average of 226 frost-free days (the "growing season"), although low-lying pockets have been known to experience freezing as late as July. Because much of the precipitation occurs during the winter months (70 percent from October through March), farms usually experience

Climate Observations

Farmers always take an active interest in the climate, and the earliest EuroAmerican accounts record the weather. Post Trader Charles Griffin began the entry of each day in his *Post Journal* with an observation about the weather, pertinent to the manager of a farm that also depended upon a transportation link with Victoria via an open strait. There he noted wind strength and direction; clear, cloudy, foggy or overcast skies; precipitation; and relative temperature. In addition, he recorded extraordinary weather worthy of note. For instance, in December 1858, he wrote that the winter was as harsh as the one he had experienced in 1853. In the summer of 1859, around the time of the Pig Incident, the weather was "oppressively hot," while on December 10, 1860, Griffin woke up to ice a half inch thick. Thunder and lightning storms, a relative rarity in the San Juans today, were remarked upon at least annually. Finally, he noted extraordinary phenomena such as the Donati Comet (September 29 and October 1, 1858), the eclipse of the sun visible at sunrise (July 18, 1860), and an earthquake "felt all over the Isld" (June 23, 1859).

drought conditions during the summer, favoring either crops that need little water, farming in water-retentive soils, or irrigation (of which there was little until after World War II). In the islands, for instance, only one hay crop is normally harvested during the growing season, in contrast to the two to three hay crops harvested in the growing season on the mainland.

> ### The Climate of San Juan County
>
> *The weather bulletin of the Climatological Service, division of Washington, contains a full page report on the climatology of Olga, compiled from the very complete data furnished by Richard C. Willis, who has been the official weather observer there continuously since the establishment of the station in January, 1890.*
>
> *The station is located on Woodside farm, fronting to the eastward on Rosario strait near Obstruction pass. It is in latitude 48 deg. 36 min. 16 sec. north, longitude 122 deg. 38 min. 36 sec. west, at an elevation somewhere between 50 and 60 feet.*
>
> *The maximum and minimum thermometers, which are of the standard Weather Bureau pattern, are exposed on a veranda on the north side of the dwelling, and are about five feet above the ground. The rain gauge is in an open space near by, and its top is three feet above the surface of the ground.*
>
> — *San Juan Islander*, August 17, 1907

Prevailing winds are generally from the south or southeast in the summer and west or northwest in the winter. However, the combination of mountain ranges to the east and south and sea-level passages between the Puget Sound and the Pacific Ocean contribute to varying wind conditions in the county. One of the more prominent exceptions is wintertime northeasters that bring frigid air down from the Fraser River Valley and over the islands. Wind conditions in specific locales may also vary depending upon the nearby topography of hills, valleys, and waterfront.

Coast Salish Farming

The Coast Salish peoples have been coming to the San Juan Islands for thousands of years to fish, hunt, and gather.

Coast Salish women maintained camas—both small, or purple, and great varieties—in beds; cultivation rights were through matrilineal lines. Camas bulbs are found naturally in areas of well-drained loam, such as island prairies, meadows, and grassy bluffs, as well as rock outcroppings with pockets of soil. The women used fieldstones to delineate the beds, which were weeded of grass and death, or poison, camas, and seeded with smaller bulbs and seeds while harvesting. The primary tool was a digging stick, a fire-hardened piece of ironwood sharpened at one end, with a handle—sometimes carved out of deer horn and then decorated—at the other. The Coast Salish harvested the bulbs in the spring (April or May) and then baked them with seaweed and herbs in rock-lined pits in the ground. They also cultivated several other edible roots, often in or near camas beds, such as chocolate lily or rice root, Columbia or tiger lily, harvest or crown brodiaea, and wild or Indian carrot.

The Coast Salish rapidly adopted the potato when it was introduced into the region via EuroAmerican trading posts in the early nineteenth century (via Fort Astoria, est. 1811, and Fort Langley, est. 1827)—or possibly earlier through indirect trade with California or Mexico. Coast Salish women continued as the main cultivators of root crops, using delineated beds and stick cultivation and harvesting techniques similar to camas production. The Coast Salish burned the savannahlike areas of prairie grass and Garry oaks to keep

Salish Woman Digging Roots

them open for cultivation of camas and other lilaceous bulbs as well as for ease of hunting game such as deer.

> *You dig a hole about two feet deep and about four feet across. In this you lay fine dry wood, then heavy sticks parallel across it, then rocks across the heavy sticks. Now light the fire. When the rocks get red hot this means get ready. When the rocks drop down, take the ashes out and level off the ground with a good hard stick. Then lay on kelp blades, salal branches, sword ferns, madrona bark, and the camas. With the camas, put all sorts of sweet bushes to infect it. The madrona bark and alder bark make it red. You must fix it so that no dirt gets in and yet leave it all full of holes. Leave a hole at the top and when it is all covered pour in more than a bucket of fresh water. When the water seeps through the rocks, it steams up. Put grass on top, then about four inches of dirt, then build a fire on top of that. Leave it all night until the next afternoon. After steaming the bulbs have to be dried a little before storing so they won't spoil.*
>
> — Saanich woman to Wayne Prescott Suttles
> *Economic Life of the Coast Salish of Haro and Rosario Straits*

Woolly dogs, which appear to have been a breed separate from Coast Salish village hunting dogs, were raised specially for their pelage. Women would shear the hair of the dogs with shell razors (later metal knives) and combine it with fireweed tufts, mountain-goat hair (from the mainland), and other materials for weaving blankets. The dogs were sometimes sequestered on smaller islands, away from the village dogs, to keep their breed pure. Coast Salish looms and weaving techniques were culturally distinctive. When the Hudson's Bay Company came into the area, workers introduced woolen blankets as trade for salmon and other native goods. After the introduction of EuroAmerican woolens and sheep, Coast Salish women gradually abandoned woolly-dog culture and adapted to both spinning and weaving sheep wool or using manufactured woolen fabric and clothing.

"Clal-lum Women Weaving a Blanket" by Paul Kane, Showing Woolly Dog

The Coast Salish did not altogether disappear from the islands after the arrival and settlement of EuroAmericans. They continued to come seasonally to hunt, gather, and cultivate camas—Belle Vue Sheep Farm manager Griffin mentions that some "Klalams" came in May of 1854 to gather "Kamass" (camas)—but because of increasing incursions of EuroAmericans on their favored camas fields, they were forced to retreat to marginal (rocky) areas or remote islands unused by EuroAmerican grazing animals. In the newcomers' description, camas beds were located in rocky areas or on islands such as Henry or Stuart. Many settlers—both employees of the Hudson's Bay Company and Americans—married Coast Salish women, who brought their traditional agricultural skills to homesteading.

> They have a peculiar breed of small dogs with long hair of a brownish black and a clear white. These dogs are bred for clothing purposes. The hair is cut off with a knife and mixed with goosedown and a little white earth, with a view to curing the feathers.
>
> — Paul Kane, *Wanderings of an Artist among the Indians of North America*, 1859

Belle Vue Sheep Farm

The introduction of EuroAmerican methods of farming—pasturage and cultivation by means of plowing, tilling, and seeding—to the islands began on December 15, 1853, when Hudson's Bay Company Chief Factor James Douglas, Clerk and later the company's Chief Trader Charles Griffin, and a company of men consisting of a mix of Europeans, Hawaiians, Metis, and Indigenous peoples, established Belle Vue Sheep Farm on the southern end of San Juan Island. They landed livestock brought from Fort Nisqually, including "1,369 sheep...1 horse, 1 stallion, 1 mare, 2 cows with calves, 1 heifer, 1 boar, and 1 sow with young," according to Douglas's report to his superiors.

Unfortunately, little is known about the man who was first Clerk and later Chief Trader in charge of Belle Vue Sheep Farm: Charles John Griffin. Griffin was born ca. 1827 in Limerick, Ireland, and raised in Montreal (known then as Lower Canada). After working as an apprentice at several locations for Hudson's Bay, he arrived at Fort Victoria in 1853. By 1856, he had been in the employ of The Company for eight years; a year later he was appointed Chief Trader, with an accompanying rise in salary. In ad-

Belle Vue Sheep Farm

dition to the principal task of managing the operations of the farm, he was required, as were all heads of Hudson's Bay posts, to keep a journal (known as the *Post Journal*) of daily occurrences, including the weather, how the laborers were employed, and special events that might impact the operations (such as the arrival of boats, disturbances from Indigenous peoples and other visitors, etc.).

In the *Post Journals*, Griffin commonly uses the phrase "Men & Inds variously employed." "Men" refers to a wide-ranging group of employees of various origins. Europeans included Scots, Englishmen, French Canadians, and Norwegians; the Hudson's Bay Company had absorbed both Scottish and French Canadian employees when it merged with the Northwest Company in 1821. Several of the Belle Vue shepherds came from the Western Hebrides. The term "men" also referred to Metis ("half-breeds") and Indigenous peoples from farther east (e.g., Iroquois) as well as Kanakas from the then Sandwich Islands (known later as Hawaii). Among the half dozen or so Kanakas employed at Belle Vue Sheep Farm was Friday, whose residence as a shepherd on the east side of San Juan Island eventually led to the place name of Friday Harbor.

Photographs taken by Northwest Boundary Survey ca. 1859

"Inds," on the other hand, referred to members of regional native groups. In the *Post Journals*, these employees were most often named for their cultural group, such as "Chimsiams" (Tsimshians), "Cowitchins" (Cowichans), "Hyders" (Haidas), "Klalams" (Clallams), "Skatchets" (Skagits), "Sneehomish" (Snohomish), and

The Company
The Very Model of a Modern Multinational

Belle Vue Sheep Farm operated within the culture of the Hudson's Bay Company ("The Company"), and their corporate practices were codified well before the establishment of the farm. Hudson's Bay was a joint-stock company (i.e., owned by shareholders), which held a general meeting every year to elect a Governor and committee to oversee its business. Those who worked for (or were "engaged by") the Company were called "Servants," who were divided into Gentlemen and Labourers. The rank-and-file was organized in a quasi-military structure: a Chief Factor and his officers (Chief Traders) commanded each trading post, while Clerks (the equivalent of noncommissioned officers) and Labourers, Shepherds, and Voyageurs (enlisted soldiers) conducted the day-to-day work. To climb the ranks, a Gentleman had to first apprentice for five years to become a Clerk. A Clerk's engagement was generally for five years, with the wage rising each year: £20 the first year, £25 the second, £30 the third, £40 the fourth, and £50 the fifth. They would then be re-engaged at £100, a salary which could then rise to as high as £150. After 13–20 years as a Clerk, one could be appointed a Chief Trader (or a half-shareholder), and then hopefully attain the highest rank, Chief Factor (a full shareholder).

The Company had a standard corporate procedure for hiring Labourers: an employee was "engaged" through a contract for a specific period of time (one month, one year), under specific conditions of work and pay. Employees could then be discharged or reengaged; leaving during the unexpired term of one's engagement, however, was considered desertion.

"Songis" (Songhees). Some tribes were also named after the Hudson's Bay fort or trading post nearest their place of origin, such as Burbank Bay, "Millbank" (probably the Bella Bellas, from Fort McLoughlin on Campbell Island near Milbanke Sound), and Fort Hope (probably Upriver Halkomelem on the Upper Fraser where the fort was located).

James Douglas, Chief Factor stationed at nearby Fort Victoria (located just 14 miles across Haro Strait), had formerly overseen farming operations at Fort Vancouver (originally located in 1824 at a site named Belle Vue) and he followed many precedents set there. The setting of Belle Vue Sheep Farm was similar to the fort—open prairie land with scattered stands of Garry, or white, oak and large quantities of lilaceous bulbs such as camas. At the fort, these prairies, which were farmed, included Fort Plain and then a series of nearby prairies linked by a road and named by proximity: First Plain, Second Plain, etc.; at Belle Vue Sheep Farm, they were called Home Prairie, First Prairie, etc. And, just as the open cultivated or grazed Hudson's Bay properties on San Juan Island were eventually squatted upon by American newcomers, an earlier encroachment had occurred at Fort Vancouver, which explains in part Douglas's rapid and strong reaction to events leading up to the Pig War on San Juan Island.

> I commenced the buildings on the banks of a rivulet, in the centre of a dry elevated sheep run containing about 1500 acres of clear prairie land, besides a large extent of brush land. This land yields excellent grass and will support from 2000 to 3000 head of sheep. I have placed Mr Charles Griffin temporarily in charge of that establishment.
>
> — James Douglas to Archibald Barclay, December 27, 1853

From the "Establishment," or compound of structures near Home Prairie, the Hudson's Bay Company employees almost immediately sought out and developed more pasture for the sheep operations, until the whole island consisted of a patchwork of sheep runs. Douglas, Griffin, and a team of Cowichans built a road (known as Cowitchin Road) north past First Prairie and Port

L'Enfer ("Hell's Gate"—later transcribed as Portland Fair) to the center of the island to access Oak Prairie (San Juan Valley), named for its surrounding ring of Garry oaks. Other grazing areas, including Mountain Prairie, Winter Station, and New Station were linked and developed. At each of these areas the laborers erected a dwelling—usually a Hudson's Bay–style house—for the shepherds and a log corral, or "park," for the sheep.

> The method hitherto most successfully pursued in the management of the Farm, is a rotation of grain with occasional hoe crops, keeping the soil in good heart, by fallowing and manures, the latter operation being most commonly performed by folding the cattle upon the impoverished land.
>
> — James Douglas on Fort Vancouver, quoted in John Hussey, "The Fort Vancouver Farm"

At Belle Vue Sheep Farm, farming methods were also similar to those used at Fort Vancouver. Livestock were grazed on naturally occurring, Native American–managed open spaces, or prairies. (Although at Fort Vancouver there was some effort to improve pasturage by cultivating and then planting timothy grass, an imported hay variety, and clover, a nitrogen-fixing legume, there is

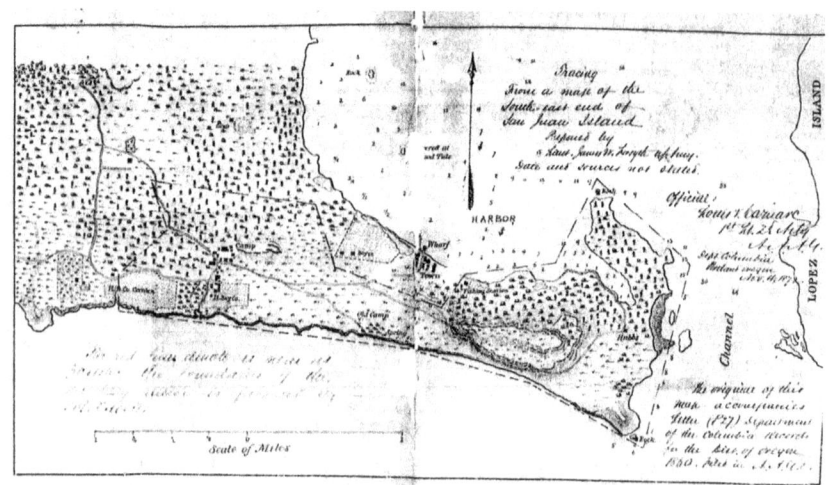

Lt. James W. Forsyth Tracing of Map of South-East End of San Juan Island

James Madison Alden Sketch of Belle Vue Sheep Farm, 1859

no evidence of that occurring on San Juan Island.) During times of shortage, laborers cut hay from swamp areas, and during the winter they fed the livestock both peas and oats. Cultivated fields were kept fertile through manuring and rotation. At Fort Vancouver, animals such as cattle and sheep were folded in fields by means of moveable fences; at Belle Vue Sheep Farm, there is some evidence of these fences, as opposed to the standard "park" or fixed fenced enclosure. Hudson's Bay Company laborers fertilized the fields with manure hauled from the barns and parks.

According to James Douglas, the crops suited for the land at Fort Vancouver were, from best to worst: corn; barley or wheat; then a rotation of peas or oats. The main crops on San Juan Island were sheep, cattle, and pigs for livestock, and potatoes, oats, peas, and turnips for field crops, with an early but short-lived experiment with wheat. Corn was apparently never attempted, no doubt due to the colder climate. Emanating from Fort Vancouver were outlying farms specializing in raising livestock (principally sheep) at Fort Nisqually (est. 1833) and grain (principally wheat) production at Cowlitz Farm (est. 1838). Upon the completion of the trade agreement with the Russian American Company and the establishment of the Puget Sound Agricultural Company in 1839, these farms grew in importance. It was from Nisqually that the original shipment of stock—composed primarily of sheep—came to Belle Vue.

The number of sheep at the Belle Vue farm grew rapidly from the initial 1,369. By January 1857, Chief Factor Douglas could report to his directors that there were 4,250 sheep; two years later both Northwest Boundary Survey member George Gibbs and the tax assessors for Whatcom County numbered Chief Trader Griffin's sheep at four thousand. This seems to be the maximum number achieved; with dwindling pasturage due to occupation of Hudson's Bay–claimed land, the numbers probably had begun to decline by the early 1860s. Like the other company farms at Cowlitz, Nisqually, and Vancouver, the principal sheep breeds included Cheviot, Leicester, and Southdown, although Griffin also records some Merino, used (as at Fort Vancouver) to improve the quality of the wool.

> I am glad to observe that you are making arrangements to part the ewes into flocks of 600 each; that is even too large a number to remain together, careful breeders generally making 500 the limit of their ewe flocks. In the lambing season, care must be taken to part the young lambs that have come during the night every morning from the flock, and to keep them apart until they are strong enough to range for themselves; shear each flock of ewes, within the lambing season, before then subdivided into two flocks each, so that you will require an additional number of hands to look after them, but the extra expense will be largely repaid by the increased number of lambs reared. The rams require much care and attention. The disease you describe as prevalent among those at San Juan is purely the effect of hardship and privation; if well fed and kept dry, the scab will soon disappear from among them; they should now receive a feed of oats daily, until they have perfectly recovered, and be well rubbed with a decoction of tobacco juice. I have spoken to McLeod about these matters; as a good shepherd ought to be well acquainted with them.
>
> — Gov. James Douglas to Chief Trader Charles Griffin, July 5, 1856

> *Owing to the carelessness of my Norwegian I have lost 13 sheep—poisoned in the same swamp where the last were poisoned—I have been out at the station all day with all the hands collecting the carcasses.*
> — Chief Trader Charles Griffin
> *Belle Vue Sheep Farm Post Journals*

Sheep operations varied little from year to year. Shepherds introduced rams to the ewes for breeding, at a ratio of about one to thirty-five, in late October or early November. Lambing would then begin around the end of March and continue through April. In April and May those male lambs that were not kept as rams would be castrated (and thus become wethers, or, as Griffin wrote in the *Post Journals*, "wedders"). In May and June, the shepherds would then wash and shear the sheep flock by flock: first the wethers, then the young ewes, old ewes, and rams. After shearing, to prevent parasites the shepherds would dip the sheep in a solution of tobacco boiled in water. In July and August, they separated lambs from their mothers (weaning), and then the cycle would begin again in the winter.

In addition to the intrinsic difficulties associated with raising sheep, Griffin's shepherds had to contend with both natural and human challenges. Several dozen sheep died as a result of poisoning, which Griffin conjectured was from the consumption of an herb located in a swamp near First Prairie. (Griffin carefully performed necropsies on all his animals that were not deliberately slaughtered in order to determine the cause of death, and his anatomical descriptions are both graphic and precise.) Wolves posed a continuing menace, and although eventually extirpated from the island through trapping, they were still killing sheep as late as December of 1859. And then there were human predators. Forty sheep were seized as a result of the tax 'sale' by Whatcom County Sheriff Ellis Barnes on March 30, 1855, and smaller numbers were shot or stolen. (In his *Farm Accounts*, Griffin attributes a total of four hundred sheep as lost to Americans.) Apparently Haidas raided Hudson's Bay Company shepherd Friday's station several times, stealing livestock.

A Typical Year at Belle Vue Sheep Farm

Because of the weather and crop characteristics, the cyclical farm year did not vary much. Plowing could commence as early as November but began in earnest after the first of the year and continued through March. Depending on the moisture in the soil, horses were used as the first draft animals, and then oxen. In March and April peas and oats were sown, while cabbage and lettuce seeds were planted, presumably in cold frames of some sort. Drills (shallow furrows) were plowed for potatoes in April, and seed potatoes cut up and sown, as were beets, carrots, and parsnips. Cabbages and celery were transplanted in May. Weeding, either through cross-plowing or hoeing, continued throughout the spring and summer. In August, cradles—scythes with long rakelike tines attached—were prepared and the oats and peas were cut and harvested into a barn or granary. Seeds were also collected at this time. In September, potatoes, then turnips, were dug and hauled to root houses. In September and October carrots were dug and onions were gathered. Presumably root crops, such as beets, as well as other kitchen garden vegetables, were harvested throughout the late summer and early fall; Griffin records digging parsnips as late as December one year. During the winter months of October through February, the grain crops were threshed and the potatoes and turnips were cleaned under the shelter of the barns, granaries, and root houses.

During the winter months, farm operations and repairs that had been put off during the growing season were dispatched. Barnyards were cleaned and the manure hauled off to the fields. Drains were installed around the establishment, particularly in the vicinity of the underground root houses (known as pits). (During the first winter—1854—it rained so much that the pits were flooded, and Griffin and his men were forced to move the potatoes underneath one of the men's houses in order to keep them dry and free from frost.) The various structures, such as the barns, granaries, and root houses, were repaired, replaced, or expanded, and fences were repaired and heightened, particularly after strong windstorms.

The sheep operation generated several products. After shearing, the wool was packed and shipped off, presumably to the company headquarters at Fort Victoria. Bell Vue Sheep Farm periodically sold off sheep as both breeding stock and meat: there are numerous *Post Journal* entries noting shipments to the other Hudson's Bay farms, to Victoria, and to the Royal Marines at English Camp.

The farm also had other livestock: horses, oxen and cattle, and pigs. The horses and oxen, although used for breeding stock, were principally used for transportation, hauling, and plowing. The cattle stock from Fort Nisqually was probably a descendant of animals that were herded or shipped from California to Fort Vancouver. What methods, if any, were used in the husbandry of these animals is not known, although there are some references to cattle and horses being periodically rounded up from various locales on the island, and both a "calf park" and stables for horses are also mentioned. Cattle and horses were also pastured at Home Prairie, as evidenced in a letter dated August 5, 1859, from Chief Factor Dallas to Governor Douglas about the damage sustained at Belle Vue upon the arrival of the American troops and the establishment of their nearby camp: "Our sheep, cattle and horses are disturbed in their pasturage, and driven from the drinking springs, in the vicinity of which the troops are encamped. (Much of the pasture is also destroyed)." There is also an obscure reference in the *Post Journals* to foals, again presumably loose on Home Prairie, being killed by US Army mules. The cattle were bred and raised for both meat and milk. Beef was a common staple of the employees' weekly rations. In 1858, the *Post Journals* record the construction of a "new" dairy, and the following spring Griffin noted with pride that there were 12 milch (milk) cows under the care of Alexander McDonald. There are also some entries suggesting sale of butter to Fort Victoria.

The Hudson's Bay Company first introduced native pigs from the Sandwich Islands to Fort George in the 1820s; later, Berkshire boars were imported from England to improve the stock. At Fort Vancouver, Kanakas were usually employed as swineherds;

although there is no direct mention in the *Post Journals*, this was probably the case at Belle Vue, because of the large number of Kanaka shepherds. There is mention of at least one pigsty, although where it was is not known (and there were no doubt more than one). The pigs were slaughtered, and the pork served, either fresh or salted, as rations for the employees, and possibly for export. There is no record of how pigs were kept or if there were any problems with poisonous weeds (such as death camas, which posed a threat when pigs were first introduced at Fort Vancouver). As for human predators, aside from the incident in which Lyman Cutlar shot a boar, there are several notations in the *Post Journals* of sows found killed around the island.

FIG. 246.—BERKSHIRE BOAR.

Period illustration of a Berkshire Boar

The farm also grew several plant crops. The largest field near the establishment—40 to 80 acres, depending on the year and the observer—was sown in oats, principally for fodder for the livestock. The smaller field was sown in turnips and peas as well as other grain crops such as barley and wheat. Oats and peas were also planted out at the Port L'Enfer and Little Mountain stations. The *Post Journals* record feeding pea straw to the sheep and cutting up turnips to feed to the oxen. In addition, on several occasions the men cut hay in swamps near Port L'Enfer to feed to the calves.

It is not clear to what extent flours or other products were made from the grain crops. Douglas wrote to Griffin on July 5, 1856, "I hope the wheat crops will turn out as productive as you at present anticipate." The following year, wheat was sent to Fort Victoria, where it was ground and then sent back to Belle Vue for consumption. There is a record of wheat being threshed in 1858, but it is not mentioned again. Douglas estimated that the average yield per acre of good land at Fort Vancouver was 20 bushels of wheat, 30 of peas, 50 of oats, and 40 of barley; on poor soils, he reckoned half these amounts. Unfortunately, the *Post Journals* do not yield enough data to compare with these figures.

Some potatoes that had been sent to Victoria in 1857 for a trial in the mess hall "excited general admiration," according to Douglas. A large field of potatoes—the seed stock having been obtained from Cowichans on Vancouver Island—was planted every year thereafter. The crop was substantial: 1,497 bushels were harvested in 1858; 800 bushels in 1859; and 1,157 bushels in 1860. Although it can be assumed that a generous portion of these were sent to Fort Victoria and other company outposts, the only known record is from the *Post Journal* of April 1859, when 300 empty bags for potatoes were sent from Fort Langley, and at least 197 returned full.

Griffin's kitchen garden included beets, cabbage, celery, lettuce, onions, and parsnips, in addition to small amounts of some of the other crops mentioned. A Fort Vancouver seed list dating from 1831 mentions several other vegetables and herbs—broccoli, cucumber, kale, leek, mustard, parsley, and radish—so these may have been also grown at Belle Vue. Griffin mentions both fencing and pruning fruit trees, but species and varieties are not indicated. He would often proudly show off his flowerbeds to visitors.

> Aleck: Angus & "Little Man" out collecting cattle, horses &c. to be seen & counted & put in inventory.
>
> — Chief Trader Charles Griffin,
> *Belle Vue Sheep Farm Post Journal*, October 17, 1859

In addition to the compound of houses at "The Establishment" near Home Prairie, Griffin and his men built several agricultural structures. To the east of the compound was the main farm building complex, consisting of one or two barns and a sheep shed, granary, shed for shearing, hen house, dairy, pigsty, and horse stable. Written records indicate the construction of a 70- or 80-foot barn (which, based on data from other Company posts, was probably 18–21 feet wide), and one of the historic photographs reveals that it had an English plan (i.e., side-entrance drive-through). All were constructed with Hudson's Bay frames; the company brought with it the men and their building techniques that they acquired when working around Hudson's Bay and the Red River Valley.

Hudson's Bay Company Frame

The general French Canadian term for log construction was *pièce sur pièce* (simplified from *pièces de bois sur pièces de bois*, meaning pieces of wood on pieces of wood). More specifically, structures that consisted of vertically grooved posts filled with planks or squared logs and the posts themselves were placed on sills, as opposed to another method in which the posts were set in the ground. With its dissemination into the Red River Valley by French Canadian voyageurs, *pièce sur pièce poteaux et pièce collisante* took on the name of Red River style. After the absorption of the Red River–based Northwest Company by the Hudson's Bay Company, the style soon became known as the Hudson's Bay Company frame, where it was used throughout the West (so much so that it is also commonly referred to as the "Canadian" style).

Although logs were used in this style of construction, they were hewn to six or seven inches square before use. A sill (*sole*) was either placed upon the ground or supported by rocks or cedar stumps. Fitted into this by means of mortise and tenon were corner and intermediary squared posts, which had a mortise of about two inches wide and three to four inches deep running their full lengths. Into these grooves, shorter logs with ends formed into tenons, or tongues, were slipped down horizontally from above, forming solid wall panels. Openings such as doorways were framed by vertical posts on either side. On the middle of the gable ends of the structures, a vertical post rose to the full height of the gable in order to carry the ridge beam, from which rafters were sloped to a plate on top of the side-wall panels and posts. Cracks between the logs were chinked with cotton, moss, or mud. Floors usually consisted of either smaller logs hewn flat or planks. Doors were constructed of planks, or slabs, and windows; where glass was available, consisted of a sash with eight- to nine-inch-square panes. Chimneys were constructed of brick hearths and stone flues, mortared with lime manufactured from clamshells as well as mud. Finally, the *Post Journals* suggest that the buildings were periodically whitewashed with lime.

An assessment of the contemporary—and potential—agricultural situation of the San Juan Islands at the height of Belle Vue Sheep Farm's operation and the beginning of American settlement can be glimpsed through the reports of the Northwest Boundary Survey (1857–1862) to the US Boundary Commission. Under Archibald Campbell, Commissioner, the team of Dr. C. B. R. Kennerly, George Gibbs, and Henry Custer surveyed the islands during the years 1859–1860. In addition to observations about current crops and farming practices of the Hudson's Bay Company, they described the fertility of the soil and predators such as wolves. Furthermore, they speculated on the economic value of the archipelago from an agricultural point of view, estimating that on San Juan Island (considered to have the greatest potential) alone, some 12,000 acres were "well adapted to cultivation," and conjectured that this could be divided into 130 "land claims" of 160 acres each (the standard American homestead allotment). The survey, completed just after the Pig Incident, set the stage for the subsequent settlement of the islands.

What about the Pig?

On June 15, 1859, an American named Lyman Cutlar shot a Hudson's Bay Company Belle Vue Sheep Farm boar that was rooting in his potato patch, located in the first prairie north of Home Prairie. Aside from precipitating the so-called Pig War, this event sheds light on the changing agricultural landscape of San Juan Island. According to the Hudson's Bay Company, Cutlar was unlawfully squatting on one of the finest sheep pastures belonging to the British government as part of the Vancouver Colony—as did the entire island. According to Cutlar and his fellow American settlers, the island belonged to the United States of America.

An American shot one of my pigs for tresspassing!!!
Charles Griffin entry in *Belle Vue Sheep Farm Post Journal,* June 15, 1859

This difference stemmed from varying interpretations of the term "through the deepest channel"—whether it was Rosario Strait to the east or Haro Strait to the west—in the 1846 Treaty of Oregon that established the 49th parallel as the land boundary between the United Kingdom and the United States.

What this incident highlights is the gradual incursion of small, subsistence farms on the island prairies that the Coast Salish had originally kept clear through selective seasonal burning and cultivation. The Hudson's Bay Company then appropriated these prairies for sheep pasture. Cutlar brought with him from his background in the Upper South/Midwest a tradition of land use that included settling on open grasslands near the edge of wooded areas while cultivating subsistence crops in the grasslands and grazing free-range livestock in the woods, where it could browse and root out native plants. He lived there with a Coast Salish wife and it was probably *her* potato patch the boar had rooted through: she brought her culture's traditional methods of cultivating potatoes that had been adapted from camas production. Although Cutlar later claimed that he had rowed to the nearest American settlement—Port Townsend—to get his seed potatoes, his wife could have just as easily gotten them from her nearby Coast Salish relations. Belle Vue Sheep Farm Chief Trader Charles Griffin, for instance, had obtained his potato stock from the Cowichans on nearby Vancouver Island. The post-shooting description of the potato patch as being imperfectly fenced on three sides, with the fourth open, matches historic descriptions of traditional Coast Salish women's camas—and later potato—plots.

The change in land use begun by the Hudson's Bay Company through their appropriation of Coast Salish camas prairies for the open-range management of livestock continued with American settlement in the islands. Where the company had grazed the fertile, open prairies, the new settlers plowed and planted crops or timothy grass and clover for pasture. Their livestock also grazed on grassy slopes and browsed in the wooded areas of the open range. The gradual replacement of prairies composed of native bunch grasses and wooded areas of either lowland alder and willow or forests of fir and pine with pastures of timothy and other hay and

> *One third of an acre in which [Cutlar] planted potatoes & partly & very imperfectly enclosed ... what has been dignified by the name of his "farm" consisted of a very small patch of potatoes, partially fenced on three sides, and entirely open on the fourth.*
>
> — Alexander Dallas, Chief Factor, Hudson's Bay Company, Letter to General Harney, May 10, 1860

grain crops effected a major change to the islands' ecosystem. Native prairie systems, which were originally dominated by perennial grasses such as Idaho fescue, California oat grass, and Junegrass, were soon transformed to a mixture of introduced perennials such as redtop, velvet grass, and Kentucky bluegrass—and annuals such as silver and early hairgrass, ripgut, downy chess or cheatgrass, soft chess, and other brome grasses.

Once the sought-after open areas were occupied, EuroAmericans cut and burned standing trees, as well as stumps and slash, to clear the land for more farming acreage. The cessation of uncontrolled wildfires or fires deliberately set by Indigenous peoples to maintain open prairies, along with the introduction of domesticated animals such as cattle, hogs, and sheep, led to the gradual diminishment of savannah/prairie land with sparse Garry oak and Rocky Mountain juniper, as well as the succession from Douglas fir forest to hemlock and spruce. Pigs, in particular, were a major factor in this change: they rooted up lilaceous bulbs such as camas and ate oak mast (acorns), leading to major changes in the Garry oak prairie habitat. This was the altered environment that EuroAmericans settlers took possession of to farm.

Dividing Up the Land

One of the requirements of the January 13, 1849, Charter of Grant to the Hudson's Bay Company of the Colony of Vancouver's Island was the establishment of a settlement of British subjects by January 13, 1854. The company's prospectus of January 24, 1849 ("Colonization of Vancouver Island") stipulated that the land, divided into parcels of no less than 20 acres, would cost £1/acre, and that purchasers of lots of 100 acres or more had to sponsor the settlement of five single men, or three married couples, for every hundred acres. This reflected the so-called Wakefield model of colonization, which assumed that sufficiently large arable areas of land would be settled in a manner replicating the landed gentry system of the homeland, England. The Hudson's Bay Company, through their local outpost, Belle Vue Sheep Farm (whose establishment in December 1853 was barely a month prior to the deadline of the five-year charter), claimed most of the land on San Juan Island as part of its farm operations. Chief Trader Charles Griffin alludes in his *Post Journal* to the fact that some distinct claims had been given to several of the employees to work on their own, but there are no extant official records of this. Northwest Boundary surveyor George Gibbs mentions that "[1] Scotchman and 2 Kanakas are said also to have taken claims for preemption"; this probably refers to claims by company employees. When interviewed by Charles Roblin in 1917, Catherine Bull (then Staff) said that "Father [John Bull] was a Hudson Bay employee and took care of sheep for Co. When he died the Co. put son in possession of property," in San Juan Valley, which she then homesteaded. The only known official document is an 1863 petition by Robert Frazer to the Colonial Office in Victoria to record a land claim; Frazer later filed a US homestead claim on the same piece of land.

In 1872, the boundary dispute between Great Britain and the United States of America over "The Disputed Islands" was settled in favor of the United States, and joint occupancy of the San Juan Islands ended. At this point, because the land was now American territory, it was subject to the public domain laws that had been applied to the Territory of Washington upon its establishment in 1853. But first, in order for public land to be "alienated" from government to private ownership, it had to be surveyed; it was not until 1874 that the islands were surveyed and divided into townships and sections. The General Land Office hired three surveyors—T. M. Reed, J. T. Sheets, and J. M. Whitworth, who began work in the autumn of 1874. They established each township and monumented section corners with wooden stakes and piles of rocks. In their field notes, they also commented on prominent

In an agricultural point of view San Juan [Island] assumes a decidedly prominent place among the rest of the islands of the Sound. Its soil is almost thoroughly good and productive, and in low situated places even rich. In the lower portion of Oak prairie, where in Winter ponds of water collect, and render the ground sufficiently moist during the Summer season, the soil is very rich and productive, its depth being from 2 ½ to 3 feet. The same can be said of some of the lower portions of the timber land where also grain of every variety could be cultivated with rich returns. According to reliable information, obtained from persons who know the island will [sic], about 50 to 60 claims of 160 acres each of good and valuable land could be laid out. Of the prairie land, about one and one half square miles is situated on the hill sides, the soil is thin and rocky, and only productive of good grass. Some portions also, those lying on the south west side of the island, are so exposed to the sweep of the southern gales that no grain or fruit could be grown there. The violence of these gales is sufficiently shown by the appearance of the trees, whose tops are bent almost at right angles to the remainder of the trunk. All land not fit for cultivation is nevertheless perfectly adapted to grazing purposes. Mr. Griffin estimates one third of the area of the island to be good arable soil, the rest only productive of grass.

— Report of Henry Custer, Assistant,
of a Reconnaissance of San Juan Island, and the Saturna Group,
Northwest Boundary Survey Commission, 1859

features (witness trees, rock formations, etc.) and at the end of each township discussed predominant natural features (timber, overall vegetation, soil quality, etc.) and potential use. Wherever there were signs of human habitation—roads, fields, fences, dwellings, outbuildings, ditches, etc.—these were also noted, with the location recorded in a number of chains (a surveying chain is 66 feet long) along the section line.

The men—and they were predominantly men (women made up less than 5 percent of those who claimed land)—who home-

The Public Land System

The United States Land Ordinance of 1785, which was first applied to the old "Northwest," in present-day Ohio, established the rectangular cadastral survey system subsequently used for most of the public lands in the central and western United States. In the Pacific Northwest, the principal Willamette Meridian, running north-south, and an east-west baseline were established near the confluence of the Columbia and Willamette Rivers in Oregon. From these lines, at six-mile intervals surveyors platted east-west township lines and north-south range lines. All of the San Juan Islands fell west of the principal meridian and north of the baseline; hence the townships enumerated from the intersections of these lines were designated Township x North and Range y West, where x and y equaled the number of six mile units from the Willamette Meridian and east-west baseline. These townships were in turn subdivided into 36 sections (or squares consisting of one mile on each side) of 640 acres each. These sections could then be subdivided into half sections of 320 acres, quarter sections of 160 acres, eighth sections of 80 acres, and so on. Because the principal lines ran north-south and east-west, subdivisions were designated as quarters of successively larger squares: for instance, the southeast quarter of the northwest quarter of the northeast quarter of Section 17 of Township 35 North Range 3 West (abbreviated as SE¼NW¼NE¼ Sec. 17 T35N R3W).

steaded the San Juan Islands came from a variety of backgrounds. Some were Englishmen or other employees of the Hudson's Bay Company (Scots, French Canadians, and Kanakas) who decided to stay and naturalize after the settlement of the boundary dispute. On January 13, 1873, 39 men, predominantly English, Irish, and Scots, but also including Austrians, Danes, Germans, Norwegians, and Swedes, applied en masse for citizenship. Many Americans had arrived in the islands after the Fraser River Gold Rush of 1858; usually, they came via Victoria from California and Oregon frontiers. Several mustered out of the US Army after serving at American Camp.

Many of these veterans were Irish or German. Of the 73 enlisted men in Captain George Pickett's Company that landed on San Juan Island in 1859, 65 were foreign born; of these, 43 were Irish and 14 German. A decade later, two-thirds (61 of 93) of the enlisted men were foreign born, and of these, over half (33) were Irish. The Irish had emigrated to America fleeing the Potato Famine of the 1840s and enlisted in the army for job security.

Because these early homesteaders were mainly single men, they often married younger Indigenous women women from local and regional tribes. The 1870 census of "The Disputed Islands," which included San Juan, Orcas, Lopez, Decatur, and Blakely Islands, was the first to record "Indians" as well as "Whites." There

> **Preemption or Homestead**
>
> Public domain in the San Juan Islands was "alienated"—deeded into private ownership—through two principal land claim processes: preemption and homesteading. In 1841, Congress passed a revised Preemption Land Act; this law was extended to the Territory of Washington in 1854, a year after its establishment. This law permitted every white male squatter over 21 years of age to claim 160 acres. To do so, the claimant had to secure a certificate from the land office as a declaration to "prove up" with a dwelling and evidence of six months residence. In addition, he had to pay $1.25 per acre in cash to secure a title to the land.
>
> The Homestead Act of 1862 allowed any head of household or single individual (including single, independent women) over 21 years of age to file for a quarter section (160 acres) of land. After filing a declaration of intent (called an "entry"), the claimant had up to six months to occupy the land. After 14 months of settlement and cultivation of the land, the claimant could purchase the property for a minimum price. However, if the claimant chose the noncash route, within seven years of the date of entry, he or she had to submit certified proof of residence and cultivation for a minimum of five years after the date of entry—a process called "proving up."

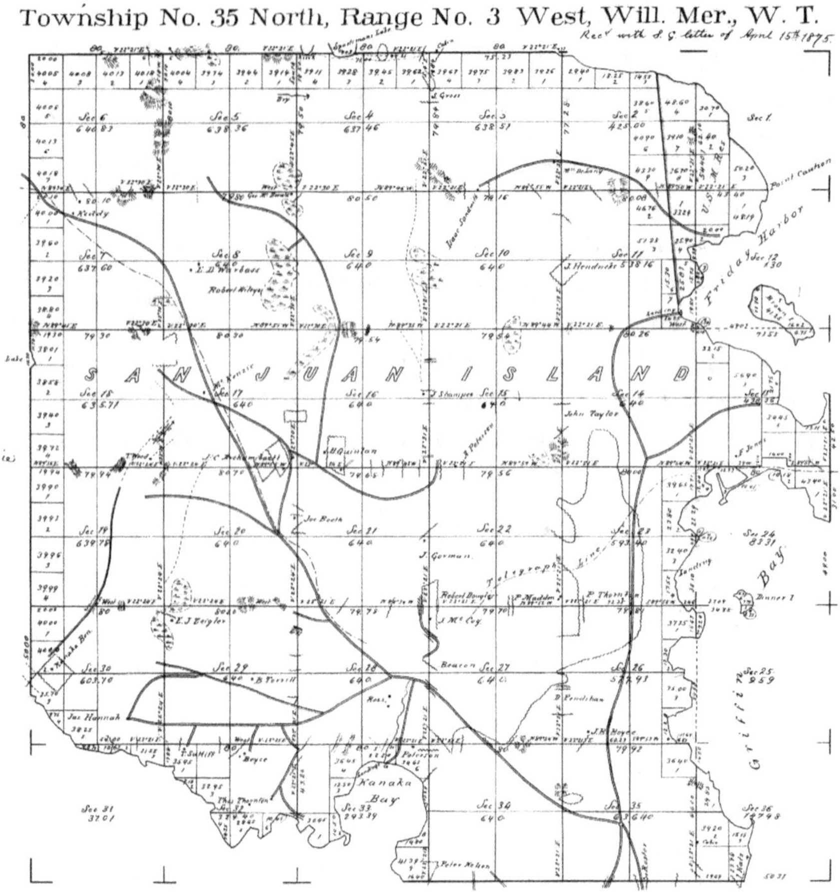

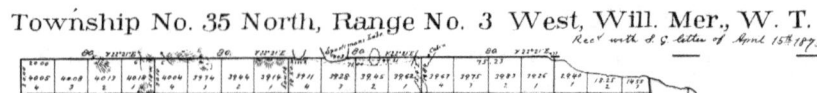

Township and Range Survey Map of San Juan Valley, San Juan Island, WA

were 189 adult men, 83 adult women, and 185 children. On San Juan Island, married men ranged in age from 21 to 60, with an average of 37.8, whereas married women ranged from 17 to 56 years of age and averaged 29–30. Of the 32 married women, 17 were native and 15 white. The native wives, however, ranged in age from 17 to 38, averaging 26.6. On the other islands—Orcas, Lopez, Decatur, and Blakely—all of the wives were natives except one. The 25 Indian wives on Orcas ranged from 16 to 35 years of age, with an average of 24. On Lopez, the 7 women varied from 24 to 30 years of age, with an average of 27. Although the overall sizes of these samples are small, most wives were younger than their husbands, and native brides were even more so—on average, about 10 years younger. Patrick Beigin, who had enlisted in the US Army around 1851, arrived at American Camp in 1859 with Captain Pickett. He mustered out of the army the following year and traded some goods for a native-made dugout canoe, which he used to transport goods from San Juan Island to Victoria and back. In Victoria Beigin, then about 30 years old, met Lucy Morris, the 13-year-old daughter of a chief of the Howcan Tribe from Sitka; they married in 1864 and homesteaded in San Juan Valley. German-born Christopher Rosler married a Tsimshian woman named Anna Pike and homesteaded near American Camp.

San Juan Valley, one of the earliest areas to be homesteaded in the islands, presents an example of the mix of French Canadian–Indigenous, Kanaka–Coast Salish, and Irish immigrants that settled here. Settlers such as J. C. Archambault, Catherine Bull, Peter Gorman, and John P. Doyle are representative of this varied heritage. James Ciprien Archambault was one of several French Canadians who came with the Hudson's Bay Company into the region, either through employment at their trading posts in the region or providing services such as fur trapping or deer hunting, which the company encouraged. Born in Quebec, Archambault arrived in the United States in 1849, and married Mary Delaunais, a Cowlitz Indian, in 1863. He naturalized as an American citizen in 1879. Although he had established residence on property in San Juan Valley as early as 1862, he did not receive his 160-acre homestead until 1883. There he built a house, barn, and other farm buildings, cultivated fields of oats, and raised sheep, cattle, hogs, and chickens.

Nearby, the land that Peter Gorman eventually farmed was homesteaded by another Irishman, Henry Quinlan, who died in 1880, aged 47, before receiving his patent. His heirs sold the estate to Joseph E. Tucker, and Tucker flipped it on October 20, 1883, to Gorman for $2,000. Peter Gorman was born in California and came with his family to the island in 1869, where his father eventually got a homestead (in 1883) in the valley to the east of Peter's

USC&GS Topographical Map of San Juan Valley, 1897

farm. Peter and his brother Joseph raised hay and grain and ran sheep, cattle, and horses on their 160-acre farm.

Next door to Archambault and Gorman was Catherine Bull, the daughter of John Bull, a Kanaka hired by the Hudson's Bay Company as a middleman, farmer, and shepherd, who married Clallam/Lummi Fu-hue-wut Mary Skqulap and settled on San Juan Island. In 1876 24-year-old Catherine, who was then listed as Catherine Emerling, applied for a patent to this family claim because she was the only one of her siblings who had not applied for a homestead. She married John Vermouth in 1880, and after receiving the land title in 1883, they sold the land to Joseph Sweeney on February 17, 1886, for $1,800. Sweeney and his wife then flipped it to John P. Doyle on March 5 of the same year.

Irishman John Patrick Doyle was born in New Brunswick, Canada, and married Mary Sweeney there in 1855; their growing family moved to San Juan Island in 1884, where he applied for citizenship the same year that he bought the farm. Doyle mainly raised sheep on the place. His son John A. Doyle bought Fir Oak Farm across the road.

Both Gorman and Doyle were part of the larger Irish diaspora in the United States—many via New Brunswick—in the 1840s. Several of their neighbors—the brothers John and Joseph Sweeney and the brothers Daniel and Patrick Madden, who gave land for the Catholic church and cemetery on the east side of the valley—were also Irishmen who came to the islands either through family connections or through their service in the US Army. Originally from Orcas, Joseph Sweeney, who had a mercantile store in Friday Harbor, often acted as a real estate broker for Irish immigrants, finding suitable farmland and extending mortgages for them to become established. His brother John Sweeney, who came from Orcas in 1890, bought the farm further down the road that John Archambault, brother of J. C., had homesteaded in 1879.

Areas of fertile land first settled by homesteaders—San Juan Valley on San Juan; Deer Harbor, Crow Valley, and Eastsound on Orcas; and Lopez Village on Lopez—feature concentrations of families that represent the rich multi-ethnic makeup of early settlers: French Canadian; Kanaka; English, Scots, and Irish; and local and regional Coast Salish and North Coast Indians.

Homesteading

Farming during the early homesteading period (1870s to 1900) was largely subsistence based. Homesteaders raised enough crops for their own domestic use and sold or traded the surplus. They would paddle, row, or sail their goods, such as milk, butter, and eggs, to markets in nearby Victoria or Bellingham. As the islands grew in population, steamboats became the main means of transportation.

The typical homestead of 160 acres or less consisted of a log cabin residence—later replaced with a one-and-a-half or two-story frame house—located on a prominent, nonarable (rocky) site and surrounded by a cluster of roughly built outbuildings: a barn, chicken house, root house, store house, and granary. A nearby small plot of land, usually 7 or 8 acres, was cultivated and planted in peas, potatoes, and turnips as well as other vegetables, while a larger field of 40 to 60 acres was cleared, grubbed, ploughed, harrowed, and then planted in timothy and clover for pasture. A small orchard of several dozen trees, usually apples and pears, was also common. Typical farm animals included chickens, hogs, cattle (used for both beef and dairy), and sheep.

> 1 dwelling, 30 by 32', four rooms, constructed mostly of drift material
> 1 good square-log barn, 32 by 32', double floor
> 1 stable, 16 by 30', shake building
> 1 woodshed, 12 by 14', shake building
> 1 wagon shed, 20 by 20', shake building
> 1 dairy, 12 by 12', shake building
> 1 root house, 12 by 18', shake building
> 1 poultry house, 10 by 12', shake building
> 25 acres under cultivation, partly timothy
> 14,565 rails in fencing (the captain thought 10,000 more accurate)
> A few 6-year-old fruit trees
> Value of improvements—$1,800
>
> — Captain G. Burton, inventory of George Jakle's homestead, 1876

One of the first acts of most homesteaders was clearing the land and preparing it for crops through cultivation. Early settlers intent upon farming crops chose land that was already clear of large trees. Some contemporary descriptions of "raw" land come from answers to questions posed in the "proving up" documents submitted by homestead and preemption claimants. For instance, according to his neighbor and witness George W. Smith, Thomas Mulno's land on San Juan Island was "Alder bottom and Fir land Clay subsoil where not Cleared, covered with Fir, Alder and Willow." On Shaw Island, August Bjork described his land as "some black sandy loam part rocky," with some "scrubly timber." Many described low ground that consisted of clay subsoil and black sandy loam covered with alder and willow as areas better for cultiva-

> *The place that Dwyer bought consisted mostly of swamp land, that is, all that was under cultivation was this meadow land, on a knoll above this meadow, sat his house, Barn, & outhouses, from his house, or, front porch one could see all over his field, which was a very pretty setting from the house. Out back of his house was some very heavy timber land, ranging up toward the mountain.*
>
> — Lila Hannah Firth, *Early Life on San Juan Island*

tion, whereas the upper, rocky areas or forests of fir and pine were identified as more suitable for grazing.

Preparing this land involved clearing all trees and brush, removing stumps, and plowing the soil. One of the early mortgages on land in the San Juans specified that the lessee was to "grub, stump, and clear and make the said lot of land thoroughly fit for cultivation." Grubbing involved the process of removing brush and ferns from the area to be cultivated, which could be accomplished either through heavy plowing with teams of horses or oxen, or by slashing and burning the brush, or a combination of both. If the land was wooded, then a far more challenging task was posed: either cutting or burning down the trees. Cutting was done either with axes or two-man crosscut ("bow") saws. Most homesteaders mention possessing among their tools one if not several axes, as well as mauls and wedges, and several list crosscut saws. Burning involved drilling two holes—one at an approximately 50° angle downward into the trunk, and another straight in to meet the first hole. Live coals were then dropped into the angled hole and fanned through a bellows applied to the second hole. Once the tree caught fire, it would burn for several days and then collapse. Sometimes the stump and roots also burned; most often they had to be pulled out with oxen or, later, when it became available, blown out with dynamite. The laborious process of grubbing would then begin. Teams of oxen, worked in pairs (called a "yoke"), pulled plows that would break up the prairie grass or fern cover and dig up any roots that were present. Homesteaders listed, among other tools, brush hooks, plows, harrows, mattocks, rakes, and hoes—all used for this process.

> *Wm. Holt gave another slashing bee last Saturday and as a result his marsh land is now all slashed. At the dance in the evening everyone seemed to enjoy themselves and tripped the light fantastic until nearly morning, despite the fact that the men had been slashing all day.*
>
> — *The Islander*, July 6, 1894

After clearing their fields, the most significant task, both in terms of size and relative importance to farming and settling, was construction of a barn. Often barns were built before families erected a more permanent home (i.e., were still living in log cabins). Most of these early barns have timber frame systems, which consist of large posts (from 8" x 10" to 12" x 16") and beams joined together with either lapping or mortise-and-tenon joints. In traditional timber frame construction, individual sections of the structure, composed of sills, posts, and braces, were framed together to form "bents" on the ground and then raised at regular intervals to form bays. Barn wrights (builders) joined the bents together with girts and plates and then added joists, collars, and rafters to form the roof structure. In early island structures, however, posts were often placed in the ground, and then joined and braced to girts to form the bays of the barn. If the girts were not long enough, particularly for the spans that covered the middle bays forming the threshing areas, they were lapped. Queen posts were then mortised and tenoned and braced at approximately one-third spans to support the purlins that in turn supported the roof rafters. This created a large open space for hay storage and grain processing: bays ranging from 12 to 15 feet on the side and heights of 15 to 18 feet to the girt, and open above, all the way to the rafters.

A Hybrid English-Coast Salish Longhouse Barn

An interesting example of what appears to be a hybrid barn, located on Blind Bay on Shaw Island, employs several cultural building traditions: the form and plan of an English barn constructed in the manner of a Coast Salish longhouse. The farmstead was originally the site of a settlement of Cowichans; later, the French Canadian hunter Julien Lawrence married a daughter of one of the families that farmed there. Hans Christensen, who applied for a homestead on this place in 1882, noted a "cedar post barn 57 x 60, good cedar shake roof $150." The extant barn, probably built during this early period, is 25' by 60', oriented lengthwise east-west, with 16' sheds on the two (north and south) long sides, thus forming an overall width of 57'—dimensions that fit the original description. The structure consists of 12"-diameter poles that have been barked and burnt; four of these poles support a beam with timber-frame, braced king posts, which, together with the 20'-high gable end posts, support the ridge beam. The plan is that of an "English" or center-drive barn, with a 12'-wide side-entry drive-through. To the west was probably the mow, where the hay was stored, while to the east was an "indent" measuring approximately 21' by 25' that, together with the drive, formed an area for threshing. To the east of this may have been storage bins for the grain. The exterior siding consists of 8"- to 12"-wide riven cedar planks, nailed vertically to the sills, girts, and plates with hand-forged nails.

Other homestead barns were less grand, and probably of standard "pole" post-and-beam construction: tall, barked and charred cedar posts (known as poles) carried beams that supported the roof, with lean-to sheds added to one or several of the sides of the barn to provide more area as well as structural stability. As opposed to later, lofted barns, the hay was piled loosely on the ground at one end of the barn, while the other areas were kept free for threshing or animals. Listed in the islands' historical records are a handful of dimensions, indicating relatively small structures: 16' x

24'; 18' x 21'; 20' x 36'; and 24' x 30'. A few are described as "cedar log," "cedar post," or "post barn with sheds." With these dimensions, assuming a height of 20 feet, the central haymow could contain 7,680–14,400 cubic feet of loose hay; at about four pounds per cubic foot and assuming a yield of two tons of hay per acre, this would imply that these early barns were adequate for hay fields ranging from 16 to 28 acres.

An important structure on the subsistence homestead was the root cellar or house, used to store root crops such as carrots, potatoes, and turnips. Belle Vue Sheep Farm had at least two cellars, or "pits" (also called "root houses"). Most likely located near the fields, these structures were dug into the ground, lined with logs, and had a log roof covered with dirt or turf. One homestead application mentions a "root house & cellar," another a "cedar log root house 9 x 12 feet," and a third a "root house 14 x 16 feet." Most extant root houses measure 9'–17' x 11'–21', or an average of 13' x 16'.

Root Cellars

The primary purpose of a root cellar was to store farm products in a uniformly cool environment, in order to keep them eatable through the winter, for home consumption, sale, and livestock feed. These included roots (beets, carrots, parsnips, potatoes, radishes, rutabagas, and turnips) and other vegetables (such as cabbages, onions, pumpkins, and squash) as well as eggs and dairy (milk, cream, butter, and cheese) and meat products (salt pork and smoked meats).

Despite the name "cellar," most structures in the islands were not fully below ground, but either embanked or partially buried into a slope or ground to take advantage of the more constant, cooler quality of the soil. Most root cellars had thick walls, some of which were insulated with sawdust, although in one instance on Shaw Island dirt was inserted between the outer (log) and inner (sawn boards) walls. In order to keep the interiors cool, ventilation was provided by means of an opening centrally located in the ceiling, with a shaft through the roof, which would draw out warm air.

Granaries

To keep vermin away from the grain, separate structures—granaries—were constructed in a tight manner with few windows and usually just one door. These buildings were elevated on wood posts, or piers, to remove the granary floor from contact with the moisture in the ground and to prevent access from mice and rats. The walls were often sided both inside and out, again to prevent animal incursions, although some structures were sided only on the inside of the structure, resulting in what is known as an "inside-out" structure, where the framing can be seen exposed to the exterior. The floors, usually consisting of a (diagonal) subfloor with tongue-and-groove or lapped flooring on top, were also tightly sealed. On the inside were either bins for the storage of loose grain or areas for stacked bags.

The capacity of a granary was calculated in bushels; each bushel contained about 1.25 cubic feet. (A bushel is equal to 4 pecks, and 1 peck equals 2 gallons). A space 10' x 10' x 6' (600 cubic feet), for instance, could hold about 480 bushels. Surviving granaries in the islands, all of which are single-story with gable roof, average from 12'–18' wide and 16'–28' long. If a typical granary were 15' wide and 22' long, with 8' high walls, its 2,640 cubic feet could contain 2,112 bushels at full capacity: an average barley yield of 45 bushels per acre meant 47 acres in cultivation; the lesser average of oats—36 bushels per acre—implied close to 59 acres planted in grain. Both figures are no doubt high, since granaries were probably not loaded to full carrying capacity.

When the Norway, or brown, rat arrived in the islands is not known, but farmers certainly needed structures to secure their grain from native, and later introduced, rodents and other vermin. Early granaries were constructed of logs—either in the Hudson's Bay Company style or a Scandinavian-derived horizontal-log construction with corner notching. Later granaries were constructed of frame lumber. The form was standard, however: rectangular in plan with a gable roof, a lack of windows or other openings except for a door, and the distinctive feature of being raised on wood piers,

Smokehouses

Tall and narrow "houses" were built to smoke meat on farms. In the days prior to refrigeration, salting or smoking were some of the few ways to preserve meat. In the smoking process, it was important to have a low-heat fire with as much smoke as possible, thereby ensuring that the meat did not cook; overheating could result in the meat getting too soft or fried, with the fats blocking absorption of the smoke. On the other extreme, because butchering was usually done in the fall or winter, it was also important to keep the meat from freezing, because smoke does not penetrate frozen meat.

In order for the smoke to rise from the ground and up and around the meat, smoke houses were small in plan (from 4' x 4' to 6' x 8') and tall (8' to 12'). They usually had only one short entrance (as little as 3' or 4' high) located in the gable end. Typically, the structure was a simple wood frame, with vertical wood siding and a gable wood-shingle roof; several racks or wire mesh shelves were located on the interior for holding the meat.

---

usually cedar or fir rounds, often several feet high. Other important outbuildings and sheds included dairies, hog houses, poultry houses, storehouses, smokehouses, stables, woodsheds, and workshops. Most outbuildings were constructed of either log or frame construction, but their form varied little in relation to use. Both stables and woodsheds could be more open, of course, but the others would inevitably be enclosed. Poultry houses, ranging in size from 10' x 10' to 12' x 16', were mainly used for chickens, although it is possible that other birds were kept there. Many homesteaders mentioned one or several chicken houses on their homesteads; Shaw homesteader Elihu Fowler had four in 1888.

The homestead was generally divided into two cultivated spaces: a kitchen garden and orchard located near the residence, and fields that were cleared and planted in a grass seed such as timothy and clover. (Later, pastures were planted with perennial grasses such as bentgrass, Kentucky bluegrass, meadow foxtail, orchardgrass, annual/Italian and perennial ryegrasses, and red and tall fescues.) The kitchen garden was used to produce subsistence crops—common vegetables as well as larger amounts of staples such as potatoes and turnips—and fruit as a cash crop, while the fields were used to pasture livestock such as cattle, hogs, and sheep. Many cultivated areas were either bottomland or former marshy areas, which had rich, productive soils that first needed to be drained to produce. Ditches were dug to drain these areas, and several homestead accounts record the number of rods of ditching, usually around 30–40; a rod is 16½ feet, totaling 495–660 feet of ditching.

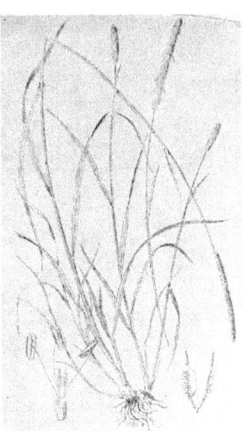

In addition to the difficulties associated with raising crops and livestock, the islands' farmers also had to contend with challenges in the form of both native and introduced flo-

Timothy Grass

Timothy was the most common grass used by early pioneers in the Northwest. It was first introduced into the area by the Hudson's Bay Company at its Fort Vancouver outpost, where pastures were often planted in timothy and clover. Timothy grows better in clayey loams than in sandy soils, and is well adapted to cool, humid climates such as the Northeast (where it was first used as a pasture crop) and the coastal regions of the Northwest. The plant grows to about 20 to 40 inches high and is topped by a distinctive furry seed head of 2 to 3 inches. Although it is often too rough to be eaten in its growing state, timothy dries well and makes excellent hay. It can be seeded in the fall or the spring; however, less seed (about 3 to 5 pounds per acre) is required at the former time than the latter (10 pounds per acre). The 1885 probate of the estate of Shaw Island homesteader Hugo Park listed about 800 pounds of timothy seed in sacks, enough to seed 160 acres in the fall or 80 acres in the spring sowing.

ra and fauna. In their encounter with a new environment, settlers had attitudes and approaches that differed from those of native peoples: one culture's resource may be another's weed. The nettle, for instance, was a significant resource plant for the Coast Salish, used for medicines and dyes as well as string and chord making, while it was considered a weed by EuroAmericans. Furthermore, along with introducing "beneficial" species, newcomers often brought weeds, pests, and predators. The simple act of bringing grain crops to the islands introduced a host of weeds inadvertently included in the seed.

> *Wild mustard has got the start of the farmers in a few instances but most of the places are free from it and the crops all through the valley never looked better or promised a heavier yield.*
>
> —*"A Trip among the Farms of San Juan Island,"*
> *San Juan Islander,* July 1, 1910

By the time the US Exploring Expedition reached the Northwest in 1840, for instance, naturalist Charles Pickering noted prostate knotweed and common lambsquarters and he names another nonnative species encountered by someone else: buckhorn plantain. He goes on to mention several other weeds (in addition to cultivated crops) associated with Hudson's Bay Company posts: knotweed, mayweed, pigweed, shepherd's purse, annual sowthistle, and low speargrass. Historian Richard White has documented the early post-contact introduction of European foxtails, American black nightshade, and dock, or sorrel, to nearby Island County. However, perhaps the introduction with the greatest effect on agriculture was the Canadian thistle. Thistle, which propagates through both seeds (the "down") and roots—particularly when they are cut up and spread through plowing—probably reached Whidbey Island by 1856. Introduction to the San Juans most likely came soon thereafter. Two plants were first introduced by Englishmen as ornamentals—Scotch broom and common hawthorn (different from the native black hawthorn): the former in Victoria in 1850, and the latter (allegedly) by Alfred Douglas to San Juan Valley for hedgerows in the 1890s. (Another possible hawthorn source could have been its use as rootstock for fruit trees.) Of the 970 vascular taxa reported in San Juan County, it has been estimated that 350, or approximately one-third, were introduced, and are mostly of European origin.

Native animal predators that affected farming included wolves, mink, bears, and deer. Wolves were a continuing menace to livestock in the early days of EuroAmerican settlement, and although eventually extirpated from the island through trapping, they were killing sheep as late as December of 1859. Both Hudson's Bay Company employees and American settlers tried to kill as many as possible, and Thomas Fleming claims that the last wolf on San Juan was killed by the company servant John Bull, a Kanaka. Several homesteaders mention the loss of their chickens to mink. Although bears are noted as being present on several of the islands, they were not necessarily considered a threat, although they were hunted and eventually extirpated. Early settlers hunted deer for meat, so they did not pose a threat to crops until orchardists began planting fruit trees in the 1880s and 1890s.

Several farms established during this time are typical of this era of homesteading. Thomas Mulno first farmed the land on Mulno Cove on San Juan Island in 1882 when he was 51 years old. In the spring of 1883, he brought from Port Gamble, where he had been working, his wife Amanda Clark, as well as his ad-

> Upon this island [Lopez] alone of the entire group did we find any positive evidence of the existence of beasts of prey. Wolves are numerous and of the larger species known to exist on our continent. Why they should be found here and not on Orcas and the other islands of the archipelago is a mystery which must for the present remain unsolved. Formerly there were a few of these animals found on San Juan, but in a very short time after its occupation by white men they almost entirely disappeared, and are now no longer any annoyance whatever to flocks. So it will be on Lopez after a few persons have taken up their abode upon it.
>
> — Dr. Caleb Kennerly, Northwest Boundary Survey

opted 30-year-old daughter, Annie, and her 40-year-old husband Alex ("A. F.") Ackley. The Ackley family already included three children—Thomas, Jesse, and Robert—and eventually grew to five with the birth of two more daughters at the farm, Abbie and Pearl. An additional member of the household was Ackley's younger brother Seward. The Mulnos and the Ackleys were originally

> *Dear sister I must describe the mink to you the mink is a small animal like a stoat or a pole cat ferret but a little larger they get in the fowele[sic] houses at night and they will kill every one if they are not disturbed they only suck the blood and then leave them I had sixty killed in one night I lay wait for him and shot him.*
>
> — Former British Royal Marine turned farmer Charles Whitlock to his sister, February 14, 1869

from Maine (East Machias and Cutler); that they came to San Juan Island via Port Gamble is significant. In 1853, Captain William Talbot, together with Cyrus Walker and Andrew J. Pope of East Machias, Maine, had established a sawmill (Pope and Talbot) and a settlement at Port Gamble. Many Maine residents, particularly from East Machias and the surrounding region, were recruited by Pope and Talbot to work at the mill, and later established themselves on homesteads elsewhere, including the San Juan Islands.

In 1883, when he filed his Testimony of Claimant upon application for a homestead, in response to the question "Was the land occupied by any other person when you made such settlement?" Thomas Mulno answered: "Yes. Israel Katz. I bought his possessory right and paid him $30." Israel Katz had established a store in San Juan Town, the small settlement of stores, hotels, and brothels that grew along the shores of Griffin Bay near American Camp. In addition to offering merchandise through the "mother" store of Waterman and Katz in Port Townsend, Katz also provided rudimentary financial services in the form of credit and loans and he brokered several land transactions on the island. Katz acted as an agent for prospective homesteaders, locating available land and charging a "finder's fee" for his services.

Mulno applied to the government for 168.68 acres, 160 of which would fall under the Homestead Act of 1865; he paid $1.25/acre, or $16.00 total, for the remaining 8.68 acres. According to his neighbor and witness to the homestead application George W. Smith, the land was "Alder bottom and Fir land Clay subsoil where not Cleared, covered with Fir, Alder and Willow." In his field notes for the 1874 survey of the section and meander lines of Township 35 North Range 3 West, Deputy Surveyor John M. Whitworth had noted "Land rolling, soil in alder & fern growth, good/balance, rocky & poor. Timber scattering fir & alder, undergrowth fir & pine, alder, willow, scotch pine, salal, gooseberry & fern." Like most homesteaders, Mulno set about clearing some land and planting it in pasture. He described the results of those first five years of work as "60 acres fenced with rails and balance[sic] with brush fence. 50 acres slashed 7 acres under Cultivation." Like most farmers homesteading a new piece of land, Mulno had a full set of farm implements, including a cultivator, harrow, two hoes, two spades, a shovel, a pick, two mattocks, a brush hook and scythe, a crosscut saw, a maul and wedges, and hay scythes and rakes.

According to his own testimony in 1889, Mulno had built two houses on the place: one in December of 1882 and the other in March of 1883. Although he does not describe the first house, it was typical of a homesteader to erect some kind of smaller shelter for the initial claim—log cabins averaged around 384 square feet (16' x 24' was typical) and consisted of one or two rooms—and then build a more substantial structure for his family. Mulno described the later house as: "Frame Lumber house 22x30, 1½ stories, 7 rooms, 8 doors + 10 windows, house ceiled throughout," and he valued it at $500. The interior furnishings consisted of "Working Range + utensils, 3 tables and tableware, 13 chairs, 2 lounges, 3 beds, cubboard [sic], 5 lamps, 4 clocks, one bedroom set." He went on to describe the other improvements as: "Post Barn with sheds, Lumber Chicken house, and a root house and hog house."

Mulno and Ackley worked the place steadily for the next twenty years, running a small, diversified farmstead. At the time of Mulno's homestead application, his livestock consisted of "6 head

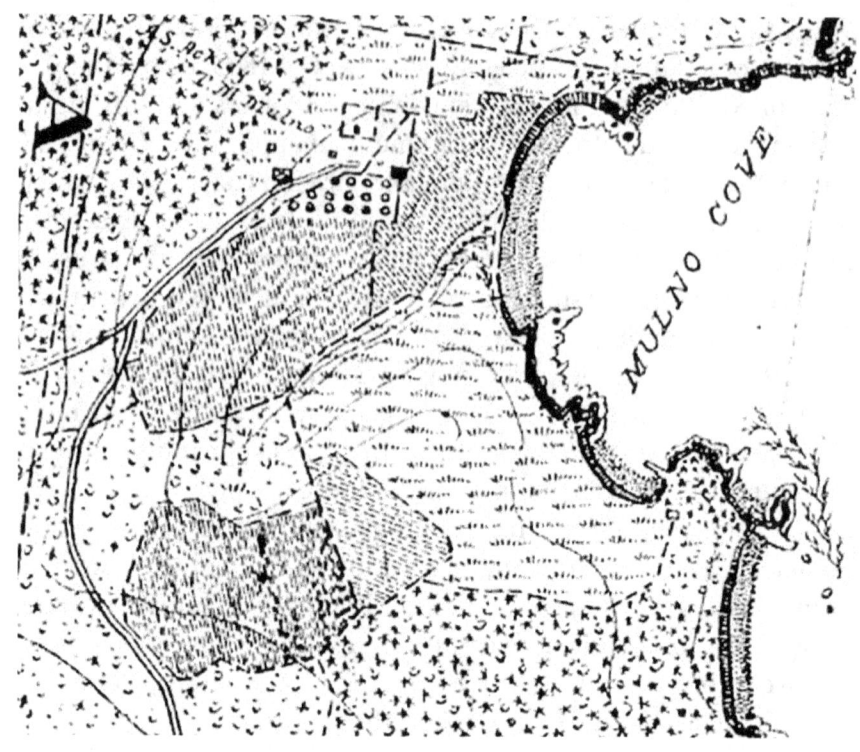

USC&GS Topographical Map of Mulno Cove, San Juan Island, 1897

of cattle, 18 sheep, 4 hogs, 40 chickens." Tax assessment sheets from the late 1880s until the year of his death (in 1902) also mention two horses, while the number of other livestock varied over the years, with a flock of as many as 40 sheep and the cattle and hogs falling as low as two and one, respectively. In addition to pasturage, Mulno also hayed his fields, which had been seeded in timothy; in 1888, for instance, he harvested seven tons. His garden was planted in potatoes, peas, and turnips, as well as "garden vegetables." After Mulno's death, A. F. Ackley continued to farm the property in much the same manner.

On Shaw Island, the 26-year-old Norwegian Theodore Tharald also claimed a 152.95-acre homestead in 1883; like Thomas Mulno, he too worked at Port Gamble to make money for developing his farm. Tharald was joined two years later by his 31-year-old brother, Thomas. (These Norwegian "bachelor brothers" were not atypical: in 1883, the Shaw Island population consisted of twelve men, all but two of them bachelors, and mostly Scandinavians.) Together the Tharalds built a log house, barn, chicken house, picket fencing (for their orchard), and split-rail fencing. Among their original livestock were 12 sheep and a cow and a calf. The brothers' crops expanded to include a large flock of sheep, hay and grains, a large vegetable garden, and an orchard of apples, cherries, plums, and pears. Their farm was located on Parks Bay, which is very close to San Juan Island, and a relatively easy row to Friday Harbor.

On Lopez Island, 49-year old James Cousins and his family homesteaded their place in 1883 at the urging of William Graham, a relative who had settled at Richardson in 1877 at the age of 35. The Graham family network is typical of early settlement in the islands: William, who was born in Ireland, emigrated to Ontario in 1841 and eventually came to Lopez Island upon advice of his relatives the Humphreys (of nearby Humphrey Head). In addition to his brother, Thomas, the Grahams also worked with N. P. Hodgson, William's son-in-law, and Albion ("Ab") K. Ridley, who was married to Elizabeth Graham. James's brother John, who moved there around 1882, was married to Ellen Burt, and her siblings Joseph and John Burt, who moved to Lopez in the 1900s, built many of the barns and farmhouses on Lopez Island. Originally James Cousins had a house built on another site on the farm, and then moved it to its present location in 1898, eventually adding a barn, machine shed, milk house, smokehouse, and root cellar to the rocky rise of land at the edge of the trees that overlooked the fields below.

From 1875, when the survey of the islands was finally completed and claims could be entered, until the 1920s, when there was no more public land available, some 749 preemptions and homesteads were filed in the San Juan Islands, with the majority occurring between 1890 and 1894. The areas of greatest potential—around American Camp and San Juan, Beaverton, and West Valleys on San Juan Island; Deer Bay, Crow Valley, Eastsound, Olga, and Doe Bay on Orcas Island, and Center Valley on Lopez Island—were the earliest to be claimed; gradually the other, less fertile lands were occupied. This process allowed for the unclaimed areas to be used as "open range."

Open Range

Because settlers first chose to homestead in areas that were already cleared and easily cultivated, rocky or heavily wooded portions of the islands were available as "open range" for grazing. While this allowed for running livestock on the public domain without private ownership of the land, livestock men had to be able to distinguish their livestock from others'—otherwise, they could not prove that the animals were theirs. An important method of identifying livestock was through the process of marks and brands. One of the earliest tasks in newly formed San Juan County (1873) was for the auditor, Edward D. Warbass, to record these proprietary markings for cattle, sheep, hogs, and horses. He wrote down a long list of filers with their distinguishing marks of their ears—crops and half crops, swallow tails, underbits, splits, holes, and even one "cut to a point like a Bull Dog"—and brands (mostly the first letter of the owner's last name), described in writing and illustrated through drawings (and even, in one case, by means of a life-size cut-out). To illustrate the significance of open-range stock raising during this period, consider the surge in the initial number of recorded marks and brands: beginning with three entries in 1873 (when the branding list had only started in December), the pages filled with 37 entries in 1874, followed by 13 in 1875, 16 in 1876, and 13 in 1877 (and then only four or so a year thereafter).

One of the ways to control open-range sheep was by taking advantage of the unique geography of the archipelago: the use of small islands for sequestering rams. The names of "Sheep Island" and "Ram Island" (located between Lopez and Decatur) stem from this use. The surveyors of the 1874 township and range survey noted several islands that were used exclusively for sheep: Dinner

> I am sorry to tell you that I had a verry bad misfortune since you last heard from mee it nearly broke mee down the cattle broke my fence and destroyed all my oates and wheat and turnips and carrots and about 10 tons of potatoes.
>
> — Former British Royal Marine turned farmer Charles Whitlock to his sister, February 14, 1869

> That it is hereby made the duty of the county auditor of each county in this territory, on application of any person residing in his county, to record a description of the marks or brands which said person may be desirous of marking or branding his horses, cattle, sheep and hogs, but the same description shall not be recorded for more than one resident of the same county.
> —An Act Relative to Marks and Brands, Legislative Assembly of the Territory of Washington, January 31, 1855

Island ("It is claimed by Fred Jones, who makes use of it for pasturage for 'Rams' for which it is well adapted, as it is covered with rank growth of bunch grass.") and Brown Island, both nearest San Juan Island. The main flock of ewes and wethers were kept on larger islands, and the rams were rowed over during breeding season. The Coast Salish used the same strategy for sequestering woolly dogs; to keep their animals' lines pure, upper-class women kept their packs on smaller islands, canoeing out to feed them every day.

> James M. Fleming claims as his sheep mark Letter (Z) size about one square inch [formed?] on the side of the face with a Brand made of Iron Right Ear Cropt and a small hole through it and left ear split.
> —San Juan County Marks and Brands, December 17, 1873

Open range, however, soon became a disputed commodity in the islands. Legally, it was common in the history of most western territories and states that those who had animals on the range took precedence over farmers who were trying to establish crops, because of the scattered nature of settlement (most of the land was literally "open" range). It was only after a major change in frontier conditions—denser settlement—that laws were established requiring stockmen to fence animals in, as opposed to the farmers protecting their crops by fencing animals out. On November 12, 1875, the Act to Regulate the Running of Sheep at Large in San Juan

County was approved by the Territorial Legislature. Stock raisers appeared before the San Juan County Commissioners soon thereafter to petition for open range. An attempt was made to repeal this law during the next session, in 1877, but range laws continued on the books.

> The great herds of sheep naturally became a nuisance after the island was surveyed and homesteaded by white families and the old range claims were at once denied. So we called a meeting and raising the necessary funds sent one of our number to Olympia to lay the matter before the legislature. The result was a stock law requiring all sheep to be enclosed and as soon as the law came into effect we proceeded to enforce it.
>
> — James Francis Tulloch, *The Diary of James Francis Tulloch*

The principal means of controlling livestock was through fencing. As early as January 29, 1855, the Legislative Assembly of the Territory of Washington enacted the Act in Relation to Fences and Fence Viewers specifying what constituted a "legal and sufficient" fence, and then regulating that "If any domestic animal or animals break into an enclosure the persons so injured thereby, shall recover of the owner of said animal or animals, the amount of damage, if it shall appear that the fence through which animal or animals broke was lawful; but not otherwise." In the next few decades, the legislature enacted a slew of fencing laws. In the Act Concerning Fences of 1869, for instance, three basic types of fences were considered as lawful in the Territory of Washington: post and rail (with either planks or rails); worm; and ditch and pole (or board or rail).

> Petition of James Peers and J. Anderson, residents of San Juan County praying for a permit to run sheep at large in San Juan County. The parties having shown that they have complied with the Naturalization Laws of the States. It is hereby ordered the Auditor issue a permit for the same in pursuance of the Act passed by the Legislative Assembly in regard to running sheep in San Juan County.
>
> — San Juan County Commissioners, *Journal*, February 7, 1876

> Wherever a farm may be located, or whatever may be its production, fence, fence, fence, is the first, the intermediate, and the last consideration in the whole routine of the operations of the farm.
>
> — Edward Todd, *Young Farmer's Manual*, 1860

The farmstead as it developed in the islands followed the formal pattern of many farms in the western United States, where different fence types were used for different applications. Often this resulted in three distinct fenced spaces: closest to the house, a yard surrounded by a picket fence; beyond that, barn yards and paddocks with post-and-board fences for the close confining of livestock; and then the ubiquitous worm fence surrounding the fields.

The first type of fence built in the San Juan Islands consisted of split cedar rails alternately stacked in a zigzag pattern that earned the nickname "worm," or "snake," fence. Local wood was plentiful and needed to be cleared to create cropland. Cedar was chosen for fence rails because of its durability—it has a high resistance to rot—as well as the ease with which it can be split. Worm fences were usually constructed 10 rails high, with two "stakes," or rails, driven into the ground at an angle to secure the corners (the rails that were then laid on top of these stakes were called "riders"). If the fence bed were 5 feet wide, and each panel of rails 10 feet long, then the linear distance covered would be 8 feet.

Before the availability of sawmills, farmers would spend winter months splitting logs into manageably sized and weighted fence-building material. According to mid-nineteenth-century sources, splitting 100 rails was a good day's work, while "a strong farmer" could split 200 a day. Diaries and journals of island farmers provide clear testimony to the many hours of labor spent on fences. At an average of 200 rails per day, it would take 75 days to enclose 160 acres, so fencing took up a major portion of a farmer's time and labor. It took about two miles of zigzag fence to enclose 160 acres, the standard farm allotment under the Homestead Act of 1862. This meant approximately 15,000 rails, with 10 rails per 10-foot panel. A two-person crew could build about 100 panels per day (as opposed to four people completing 35 to 40 sections of post-and-rail fence).

> [Worm fences] shall have not less than four feet worm to rails of ten feet in length (and if greater length, in that proportion,) shall be four feet high, well staked and ridered upon that, according to practice. Below the third rail from the ground, no crack or space of more than five inches shall intervene, and below two feet in height, there shall be no crack or intervening space more than seven inches, and the whole height of said fence shall not be less than five feet.
> — Laws of Washington, 1855

When sawmills became more prevalent, sawn lumber was used along with cedar fence posts for fencing. Often imperfect boards (with rough corners and/or some bark) were used in fencing, while the best boards were reserved for building construction. Post-and-rail fences consisted of five- to six-inch diameter posts set in the ground with five to six rails of split or milled lumber nailed horizontally. This type of fencing was most often used for paddocks and other types of livestock-holding areas.

A Breachy Cow That Led to Bloody Murder

Disputes over fencing wayward animals took a bloody turn on Lopez Island in 1882 when John Kay shot to death his neighbor John Anderson over a wandering ("breachy") milch cow. The two neighbors had a fence between their homesteads, running down to the beach, but at low tide the cow could get around it from Anderson's pasture to Kay's. When Kay corralled the animal on the afternoon of Tuesday, May 15, 1882, Anderson came to claim the cow and an altercation ensued. Kay, with the help of his wife, Eliza Jane, got the upper hand and shot Anderson in the chest at such short range that the gunpowder set his shirt on fire. Although Eliza Jane threw water to put out the flames, Anderson expired from the gunshot. Kay was tried for murder and sentenced to prison; Anderson's estate went to his wife, Lucy.

Rough, riven cedar or fir pickets were used for the taller fences surrounding orchards. These fences kept deer as well as livestock out. The pickets were usually attached to top-and-bottom rails nailed to cedar posts. In his application for a homestead in 1888, Shaw Islander Theodore Tharald noted under improvements

> *[January 1869] Wether has been fine some rain so as we could work all the mounth. I have split 1,100 rails and helped James with his fencing. [January 12, 1870] 176 split; in the month of February (wether blustrey, some fine days, little snow), 1873, some 900 rails.*
>
> — Thomas Fleming, *Journal* entries

"75 rods [a rod is 16½ feet, so 1,237 feet in total] Picket Fence $100" (as compared to his "150 rods rail fence $140"). Charles Griffin had noted in his Belle Vue Sheep Farm *Post Journals* that he had his men cut willows to put around his garden fence to keep hens and other poultry out—a technique that was derived from the older English wattle, or woven fence, consisting of pales driven into the ground and interwoven with branches or twigs. Fancier, sawn lumber pickets (often with the tops cut in diamond, lozenge, and other fancy shapes) were used later to keep livestock out of residential yards. Prevalence of these pickets depended upon the availability of milled lumber, as well as the extra money needed

Fences at the Firth Farm (American Camp)

> He [James Hannah] found a cedar tree, with good splitting qualities in it, down it came with the help of a well ground ax & was sawed into 4 foot lengths, split in strips, & rived into pickets. Posts of that fence were also split out of that tree, & in a very short time it seemed to me, Father had a picket fence all around our house, a foot board around the bottom of the fence, which when finished made the fence 5 feet high.
>
> — Lila Hannah Firth, "*Early Life on San Juan Island*"

to purchase it. Although often associated with yards in front of houses on town lots—where they also served to keep out livestock roaming in the streets—picket fences can be seen in historic photographs adorning rural farmyards as well.

Commercial Farming

As the land was divided up and settled, farmers expanded their operations beyond a subsistence-based economy. The first US census, conducted in 1860 on San Juan Island only and including the soldiers at American Camp, revealed that most settlers listed their field of occupation as agriculture: 26, or 52 percent, of the 50 civilians were farmers. An even higher percentage were enumerated in the 1870 islands-wide census of "The Disputed Islands," which noted 112, or 65 percent, of 173 total workers in the field of agriculture, be they farmers, farm laborers, shepherds, or sheep growers. The number of farms jumped from 84 in 1880 to 278 10 years later, while total farmed acreage grew from 17,572 to 41,761 during the same period. The percentage of farmers among the overall workforce remained high throughout the 1880s: at least three-quarters of those employed on Lopez and Orcas were enumerated as farmers in the 1880 federal census and the 1887 and 1889 territorial censuses, while on San Juan, always the more diversified economy, at least half identified as farmers. By 1900, other major island industries—limestone quarries and lime kilns as well as fish traps and canneries—began to employ more workers, so the percentage of farmers on each island dropped: Lopez, 47 percent;

Orcas, 53 percent, and San Juan, again, the most occupationally diversified, only 24 percent. Some of the smaller, less populated islands were still predominantly occupied by farmers: 70 percent on Shaw and 45 percent on Waldron (where the other half were fishermen). The change from subsistence-based homesteading to commercial farming is corroborated by the rise in the number of "farm laborers" (i.e., those who worked on a farm) alongside "farmers" (i.e., those who owned and operated their farms).

Most of the farmers during this period had small, diversified commercial farms. Personal property assessments taken in 1891 indicate that island farmers had an average of two to three horses—enough for a team of two "draft" animals—and three to six milch cows, with larger numbers of sheep and hogs depending upon the operation. But this period also saw the rise in large farm operations, as evidenced by newspaper coverage in such booster pieces as *The San Juan Islands, an Illustrated Supplement to the San Juan Islander* (1901) and the *Islands of San Juan County, Washington*, a special edition of the *Everett Morning Tribune* dated July 18, 1908. Eight-hundred-acre GEM Farm, established by Ben Lichtenberg on Lopez Island in 1898, produced several crops, including fruit such as apples, pears, plums, and cherries; poultry such as Plymouth Rock cockerels and White Holland turkeys; and dairy from Jersey cattle. The Orcas Fruit Company near Eastsound was

View of King Farm with Friday Harbor in the background

"GEM FARM"
Poultry and Fruit Ranch. Ben Lichtenberg, Prop.

An institution of magnificent proportions and of wide importance in the commercial and industrial circles of this county is the "Gem Farm," which is situated two miles north of Lopez and is a town all in itself. Here is embraced a ranch of over 800 acres, under cultivation and in grazing lands, than which there is no finer in the State. An orchard of 500 prolific fruit trees of the choicest selections of apples, pears, plums, cherries, etc., is one of the principal features of the place. However, especial attention is called to the chicken industry, where only the best breeds are raised, and care is taken to have them all full-blooded and of select varieties. One breeding house 20x125 feet is maintained, also broiler houses and individual houses. Three incubators, with a capacity of 500 eggs at one time, are used, also brooder houses and fine outdoor brooders, everything being according to the latest improved methods. They have now on hand from 700 to 1,000 chickens. They have sold Plymouth Rock cockerels for from $2.50 to $20 and hens $6 apiece for breeding purposes, and maintain the only breeding houses in this State. They also keep two magnificent White Holland turkey gobblers for breeding purposes—the only ones known in the Northwest.

At the Gem farm is also kept a fine drove of Jersey cows—full-blooded—and much attention is paid to dairying. The one aim of this institution is to have the best of the best and only deal in blooded cattle and chickens, and the products of this place are fast becoming recognized for their superiority.

— *The San Juan Islands, Illustrated Supplement to the San Juan Islander,* 1901

formed in 1907 by J. E. Moore from an existing ranch. Consisting of some 4,000 apple, pear, and prune trees on 60 acres, the company had its own cannery and fruit evaporator and shipped thousands of cases of fruit off-island. Bellevue Poultry Farm, es-

GEM Farm, Lopez Island

tablished around 1910 near Roche Harbor on San Juan Island by Tacoma and Roche Harbor Lime Company's John S. McMillin, specialized in Crystal White Orpington chickens as well as geese and turkeys; photographs reveal an extensive complex of poultry houses and runs occupied by thousands of birds.

The two decades bracketing the turn of the nineteenth century—1890–1910—was also a period of great agricultural infrastructure improvements, in particular, barn building. In the local

ORCAS FRUIT RANCH AND CANNERY

In March 1907, Mr. J. E. Moore acquired from Mr. Pike the latter's well known ranch at East Sound, consisting of 91 acres of which 60 acres are in fruit, principally apples, pears, and prunes, and 10 acres in meadow. There are 4,000 fruit trees in the ranch and immediately upon acquiring it Mr. Moore erected a cannery with the object primarily of dealing with his own products. Last year 2,300 cases of fruits from his own trees were so dealt with. This season, if the market conditions are favorable Mr. Moore intends to extend the scope of the cannery by buying from local growers. This is the only cannery on Orcas Island and it contains three retorts and has a capacity of 150 cases a day, it is probable that a large business will reward Mr. Moore's enterprise in this manner.

— Everett Morning Tribune, July 19, 1908

newspapers—*The Islander* (1891–1898) and its successor the *San Juan Islander* (1898–1914)—there was regular reportage on each island's farming scene ("Valley Gleanings," for example, for San Juan Valley), and barn raising was prominently featured. The year 1894 was a banner year, with 10 new barns started; this was exceeded in 1905 with 21 raised, and subsequent years ranged from 4 to 12 each year. In the twenty years from 1894 to 1914, the newspapers reported over 120 barn raisings—a significant number, given that a 2009 survey of historic (pre-1959) barns estimated only 120 extant in the islands.

After the use of oxen in the early years of EuroAmerican settlement, horses became the primary means of animal power on farms in the islands. Based on the federal census of agriculture, the average number of horses per farm in San Juan County was about two, which would suggest that most farms only had a single team of paired draft horses. The nineteenth century witnessed great

Agriculture and Fruit Raising

The soil of the valleys is very fertile, producing immense crops of grain and hay, while the upland is unexcelled for the production of fruits and vegetables and for dairying and grazing purposes. Oats, wheat, barley, rye, potatoes, timothy and clover grow in perfection and yield largely. Owing to the gentle spring rains, timothy and clover yield immensely when compared with Eastern states; and it is not unusual that a yield of three to five tons to the acre is had. Dairying is profitable, and, owing to the mild winters, cows give the most milk in the fall, winter and spring. Farmers who have turned their attention in this direction have been very successful. The great quantity of eggs and poultry shipped from the islands embraced within San Juan County speaks volumes for the poultry industry, and yet this industry is only in its infancy. Fruits of all kinds thrive here. Apples, pears, cherries, plums, prunes, strawberries and blackberries grow in perfection and in profusion.

—The San Juan Islands,
Illustrated Supplement to the San Juan Islander, 1901

> Lopez Island is rapidly gaining the reputation of possessing and breeding mighty good horse flesh. Among others, James Buchanan has a splendid stock farm and some first-rate animals. He has three Clydesdale beauties which are destined to materially improve the work-horse strain of the Sound, and a barn full of other breeds equally good in their way. Jim Davis, of Richardson, also has a fine lot of horses.
>
> —*The Islander*, March 6, 1896

improvements in horse-drawn and horse-powered machinery for agricultural tasks such as plowing, sowing, harrowing, cultivating, reaping, threshing, and winnowing, and many farms in the San Juan Islands had an extensive assortment of these machines. Although tractors were introduced to the islands as early as 1924, most farms continued to use draft horses until the end of World War II. After reaching a high of 1,102 in 1920, the number of horses and mules in the county declined in the next several decades, but it was still around 500 in 1940. Steam as a power source was used on some farms: steam threshers were hauled from farm to farm and barged from island to island, and steam engines were used in powering canneries and creameries.

Very few barns were specifically dedicated to horses in the San Juan Islands, probably because of the low number of draft teams on farms. Most general-use barns incorporated stalls for workhorses. The size and design of barns specifically used for horses depended upon the number of animals to be housed. The 1923 edition of *Louden Barn Plans*—a standard plan book—recommended 75–90 square feet of space per horse (including the alley), so a barn housing three draft teams, or six horses, might be approximately 480 square feet or 20' x 24'. The stalls themselves would be about 5' wide x 9' long, although teams could be housed in double stalls of 8' x 10'. So-called box stalls, constructed of standard dimensional planks nailed horizontally, ranged from 9' x 12' to 12' x 12'; these were used for carriage or riding horses—not draft. Many design manuals during this period recommended windows: R. A. Moore et al.'s 1920 *Farm Economy* advised "one 3' x 3' window should be

Apple and Pear Exhibit, Orcas Island

provided for each stall, and it should always be low enough for the horse to look out"; it also recommended "light, airy, and dry shelter" because ventilation was very important in horse barns.

Barns were not the only agricultural infrastructural improvements built for specific crops or livestock. In 1901 creameries were established in Friday Harbor and Eastsound, and five years later on Lopez. Several canneries were also constructed during this period: the Island Packing Company in Friday Harbor (1894), which canned fruits and vegetables as well as seafood; the cannery built in 1907 by J. E. Moore at the Orcas Fruit Ranch, which canned exclusively fruit; the Shaw Island Canning Company (1912), which canned sea products in addition to fruit and vegetables; and on Lopez (roughly 1906). Many fruit raisers, particularly those on Orcas, constructed barns designed to store their fruit crops as well as evaporators that would dry fruit for the market. Trading companies established warehouses at several of the principal ports—Friday Harbor on San Juan Island, and Eastsound, Orcas Landing, and West Sound on Orcas Island—for storing produce prior to shipping such as fresh vegetables; grain; fresh, dried, and canned fruit; and seafood.

In the 1870s and 1880s the Territorial Legislature passed several laws providing for the construction of ditches, dams, and watercourses. These projects allowed farmers easier access to the rich, black, bottom soil of the valleys, which retained moisture during the dry summers. In 1892 farmers in Beaverton Valley on San Juan Island successfully petitioned the San Juan County Board of Commissioners to establish a drainage district in order to drain the excessive water from the marshy land. (After falling into disrepair, in the 1920s the Beaverton Valley ditch was revitalized through the work of an association that assessed and taxed landowners along the ditch for necessary repairs and improvements.) Two years later Ditch District No. 2 was established for the Davis Slough area on Lopez Island, specifically calling for the construction of a "flood gate." In the early 1920s, the US Army Corps of Engineers worked on portions of the San Juan Valley ditch. As part of this work, the Corps and the Cooperative Extension Agent advised farmers to plant reed canary grass as a conservation measure; this grass has subsequently spread throughout the wetlands of the valley.

During the twentieth century, both the total acreage farmed and number of farmers grew each decade until reaching highs of

Draft Horse Team Hauling Stone Boat

68,513 acres in 1920 and 566 farmers in 1925. This period saw the islands' agriculture grow to commercial-scale stock raising (sheep and cattle), dairying, and cultivation of various crops such as fruit, grains and hay, and poultry (chickens and turkeys).

Sheep

Sheep, raised for breeding, meat, and wool, have been a major crop in the San Juan Islands since 1853, when Charles Griffin of the Hudson's Bay Company brought 1,369 sheep to the newly established Belle Vue Farm on the south end of San Juan Island. The company eventually built up its flock to some 4,000 animals. American and British settlers soon established large flocks. After the departure of the Hudson's Bay Company in 1862, large stock production—particularly sheep—was taken up by men such as the Keddy brothers—John on "Cady" (the American transcription) Mountain on San Juan and William at the head of Fisherman Bay on Lopez—who were British citizens, originally from Victoria. Not only did they run their own sheep, but they also let contracts to other shepherds to raise their stock and harvest the lambs (one-half the crop less loss) and wool (one-half of the crop).

Sheep and Stock

The finest grazing ranges in Western Washington are found in this county and large interests are engaged in sheep and stock raising, especially on San Juan and Lopez Islands. The ranches on Lopez are a source of pride to any county. There are more sheep in this county than in any other county in the State west of the Cascade Mountains, and the superiority of "San Juan mutton," which is the choicest found in the city markets, is attested to by the increasing demand which exists for it.

*—The San Juan Islands,
Illustrated Supplement to the San Juan Islander, 1901*

Two other early stockmen were German-born Augustus Hoffmeister, who was post sutler (civilian storekeeper) at English Camp, and Samuel Trueworthy, a native of Maine. Hoffmeister ran

Frend Bowker
> *Dear Sir*
> *I ham Sore that I hav bee the cas of wonding you feelings. But I will not be so ... a gan I hav ben a long tim tring to send sheep hover but no bot an wen I got a bot I was on the way to the Post hofes with leter the anser to Mr John tods leter so it was not poste. The Sheep that I brot over wold not Sut you for thay was lams as I had no tim. The Bot was loded with Shingls and I sold to Stafort for 3 dollers and Payd to Rup and Company 100 dollers and frat 25 dollers Duty. So you will See that I tend to pay my Dets. Now my old frend Bowker you will not hav to find any mor falt as I ham old Bill Smith agan. So I hop that haur frenship will last. It was my intencen to com with mor Sheep. But the bot went to the Squinish ... and hav loaded and cold not tak no Sheep but promist to bee back in Six days. To day I hav Sent aman to hutchinson for a bot if no bot to go to San Jun this tusty the 13 no word of bot or man and Snowing fast. I ham Sorey to hear that mrs Bowker is Sick and hop She will sun get beter. Giv my best wishes to all of the family and a hapy new year. So I must clos as I hav to go Six mils to the Pos hofes and the Sno faling fast*
>
> *Your truly — William Smith, Orcas Island, to J. S. Bowker of the firm Bowker and Tod, Victoria*
> *January 10, 1874*

both cattle (60 head) and sheep, of which some 500 were Southdowns that he had on his home ranch near the camp as well as on Spieden Island. When Hoffmeister died in 1874 (his was the first probate filed in the newly formed San Juan County), his estate was divided between Isaac Sandwith, who leased his property on San Juan, and John Tod Jr., who purchased his sheep, farm, livestock, and improvements on Spieden and Henry Islands. Trueworthy ran his sheep—numbering 800—on his ranch near Westsound on Orcas Island; according to his 1876 probate he also had some cattle and between 300 and 350 goats on the mainland.

In her memoir, Lila Hannah Firth relates that her father, James Hannah, raised free-range goats on Mount Dallas in the

Driving Sheep down Spring Street, Friday Harbor

late 1860s, after the Hudson's Bay Company had withdrawn their sheep. However, the area was later used by shepherds based on the north end of the island, and in an apparent "range war" the shepherds shot most of Hannah's goats.

One of the problems in documenting sheep and wool production in the San Juan Islands during the latter half of the nineteenth century was smuggling. In 1861, the US Congress enacted a tariff sponsored by Representative Justin Smith Morrill. The Morrill Tariff imposed a duty on the import of wool, among other items, allowing sheepmen a better price in the United States than across the border in Canada. Smugglers would ship wool in sloops at night from nearby Vancouver Island to the San Juans. Locals also used small, fast ships that could land on secluded coves under the cover of darkness to move flocks of sheep that were subject to seizure by the county sheriff, or simply scattered them among the islands to evade confiscation.

The Hudson's Bay Company started with Cheviots, Leicesters, and Southdowns, with some improvement from Merinos, at their Belle Vue Sheep Farm, and these were probably the main breeds used by subsequent sheep raisers in the islands (as noted above,

> Orcas Island being quite mountainous was largely devoted to sheep raising by the early settlers who each claimed certain portions of the island as their range. Sheep thief was quite a common term to fling at each other. But local stealings were small compared to the exploits of certain enterprising citizens of Victoria who often made forays and with the aid of trained sheep dogs rounded up whole boat loads on moonlight nights.
>
> There was a custom among the squaw men who had sheep in the early days of cutting off the tails of all lambs at round up time each spring. Anyone who had sheep on the range had the right to kill any longtailed sheep at any time. So Charlie Basford, a neighbor of ours who had bought a half dozen wethers and turned them loose boasted that he lived off the increase. I don't believe he overstated his case judging by the number of sheepskins I have seen on the mountainside from time to time. We were glad when the vexing sheep problem was settled and every man had his just allotment of sheep.
>
> — James Francis Tulloch, *The Diary of James Francis Tulloch*

Hoffmeister had Southdowns). In an article titled "Stock Raising in San Juan County," the Everett Morning Tribune reported that

Sheep on the Friday Harbor Dock Awaiting Shipment

> *The price of wool on the Canadian side being about half that paid on our side made raising wool on a foggy night extremely profitable and J. L. Sherer, our county auditor who made a careful canvas, found that the yearly yield of San Juan County was 27 lbs. while the actual clip was 2½ lbs.*
>
> — James Francis Tulloch, *The Diary of James Francis Tulloch*

most of the sheep in the islands were Shropshires, Southdowns, and Oxford Downs. It went on to note that the wool crop averaged nine pounds per head and that lambing averaged 100 percent. In the 1930s, San Juan county extension agents reported that the predominant breed was Oxford Down.

The number of sheep in San Juan County rose from 10,266 in 1880 to 12,871 in 1900 and then fell steadily in the next three agricultural censuses, undertaken in 1910, 1920, and 1925. However, in the 1930 census it jumped to the highest number recorded

> ### Andy Johnson's Big Wool Crop
>
> *Andy Johnson, whose fine farm extends down to the shore line at the south end of San Juan island and affords a convenient landing place for boats from the British side of the channel, says his wool crop this year runs about one hundred pounds to the sheep. Eric Erickson says that his sheep are doing better than that, although his farm is several miles inland from the coast. They were talking to the deputy collector of customs when they imparted this information and of course, as law abiding citizens, opposed to smugglers and all their works, they could have no desire to mislead that officer. Andy rather turns up his nose at the small yield of wool reported by M.A. Ward and mentioned in the ISLANDER last week—only 21¼ pounds in one fleece, and intimates that with the great advantage of location enjoyed by Mr. Ward at his Kanaka bay farm, his annual clip should be considerably larger.*
>
> — San Juan Islander, June 9, 1906

in the islands: 18,789. Part of the census-to-census fluctuation in numbers was due to the relative costs of dairy versus sheep production: in the late 1930s, for instance, the scarcity of good labor and the low price of dairy products induced farmers to switch to sheep, because they could get a good price for both wool and lambs with less labor.

Hay and Grains

Because of the islands' climate and lack of irrigation, only one hay crop was harvested in the growing season. Hay, both wild and tame (cultivated), was originally harvested loose (i.e. cut directly in the field), loaded into wagons, and then driven into a central bay of the barn, where it was forked into the "mows" on either side. In 1867, William Louden patented the Louden Hay Carrier, a mechanical system used to load hay into the mows. This consisted of either a rectangular wooden beam or metal track hung from the underside of the roof ridge, upon which a trolley could be wheeled to position over select areas of the haymow. Pulleys on the trolley were threaded with ropes that were attached on one end to either hay hooks or slings, and on the other to a draft team (and later a tractor or truck) to pull it, and thus elevate the load. With the load suspended high above the mow or loft, the trolley was

Lizzie Lawson Haying

wheeled into position and the load released by pulling a trip rope that would disengage the hooks or sling. After 1900, several companies provided standard equipment for mechanical hay systems. One of the most prominent was the Louden Machinery Company of Fairfield, Iowa; the company's models included the Louden Junior Hay Fork Carrier, examples of which can still be found in several island barns.

The hay rail-and-trolley system transformed barn design. Newer plans, called "western" barns, featured an entry in the gable end for a mow in the middle and stalls on either side. These were often post-and-beam construction, which created a large open space in the middle, and the mow extended from the floor (usually unfinished, i.e., dirt) all the way to the roof; sheds were built on the sides for livestock. Hay track-and-trolley systems limited the width of the barn (traditionally ranging from 20' to 36'—the most common being from 34' to 36'), because of the range of lateral dispersal of the loose hay from the central track. Maximum economy of storage could then be obtained through going up in height, although limited by the costs of the heavier construction and the danger of wind shear—tall barns basically act like large

Louden Hay Barn Diagram with Rail-and-Trolley System

Harvesting Oats, Paul Guard Farm, Beaverton Valley, San Juan Island

sails. Several new roof shapes—gambrel (English and Dutch design), Gothic arch, and bow—formed with frame roof-truss systems consisting of shorter, lighter dimensional lumber, added volume by means of lofts, allowing for a lower level to be used for animal stalls. The length of the barn would then be determined by the capacity requirements. The capacity of loose hay is about 500 cubic feet per ton, or four pounds per cubic foot. If the loft or mow of a barn was 35' wide by 50' long and 25' high, its capacity was 87.5 tons; with an average yield of two tons per acre, a barn of this size could hold the hay crop from a 40-acre field.

Grain farming developed gradually through the last quarter of the nineteenth century. Grain crops included barley, oats, wheat, and dry peas. According to the agricultural census, total county land planted in oats, with yields of 1–1½ tons per acre, reached a high of 2,104 acres in 1900. Barley and winter wheat (red Russian variety) were never as extensive, although the former peaked at 1,007 acres in 1930, while the latter peaked at 863 acres in 1959. The wetter, better drained soils of San Juan and nearby Beaverton Valleys on San Juan Island, Crow Valley on Orcas Island, and Center Valley on Lopez Island proved excellent for grain production becuse of retention of moisture during the long, dry summer season.

> *Mr. Jas. Buchanan and his force of men finished threshing for Wm. P. Holt last week. Mr. Holt had about forty-five acres in oats and he received a yield of forty-eight tons of very fine grain.*
>
> — *The Islander*, September 26, 1895

By 1886 there was enough grain production on San Juan Island to prompt a group of men from Port Townsend to invest in a mill at Argyle, a shallow water port in Griffin Bay. This three-story structure, with a basement and superstructure for accessing equipment, had a loading dock at wagon height and was powered by steam. By varying the millstones the machinery could grind grains—barley, oats and wheat—as well as other crops, such as split peas. Next to the mill building was a stone-and-brick kiln for drying various products, including prune plums. The mill closed in 1909 due to lack of business.

At Belle Vue Sheep Farm the oats and peas were cut in August with cradles (scythes with long, teethlike attachments for laying grain in bunches as it was cut) and harvested into a barn or granary. Early threshing was done in "English-plan" barns with side entries or center drives. The center bay, along with a bay off to the

Walter Burt with Horse-Drawn Binder, Lopez Island

Harvesting the Prairie at Firth Farm (American Camp)

side, provided an approximately 30-foot-square area for horses to walk in circles, treading on the shocks of grain. With both doors of the center drive open, and likely oriented towards the prevailing winds, the breeze would blow off the chaff, leaving the grain on the floor. It would then be gathered up and stored in grain bins near the threshing floor.

In 1876 John Bartlett, of Lopez, brought the first threshing machine—an eight-horsepower "Sweepstake"—to the islands. Later, Lopez farmer J. A. Buchanan bought a J. I. Case "improved" separator, which was run off a steam engine, and together with

J. A. Buchanan

In his threshing business he has built up a large range of business and travels to all the larger islands of this county. Mr. Wright owns and operates the thresher which is a J. I. Case improved separator and can be used in threshing wheat, oats, peas, flax and all kinds of grain, while Mr. Buchanan owns and operates the steam traction engine—a 10-horse power Massilin traction engine.

—*The San Juan Islands,*
Illustrated Supplement to the San Juan Islander, 1901

> **Notice to Farmers**
>
> I have purchased a fine new J. I. Case threshing machine and will thresh wheat and oats on San Juan Island for $1.00 per ton and peas for $1.50 per ton.
>
> — Wm. Buchanan, *San Juan Islander*, October 1, 1909

neighbor J. T. Wright operated it on farms throughout the islands. Steam threshers were hauled by horse teams from farm to farm and even barged among the islands. These operations consisted of a steam engine that drove a long, wide belt attached to the threshing machine, which would separate the grain from the chaff into separate piles. Grain was put in cloth (usually burlap) sacks, which were then sewn shut in the field. Threshing of grain was a community effort: crews of men from surrounding farms would meet at the fields to be threshed, and farm wives would vie with each other to provide ample lunches for the workers.

During the latter half of the twentieth century, the decline in oats and rise in barley corresponded to the gradual eclipse of the

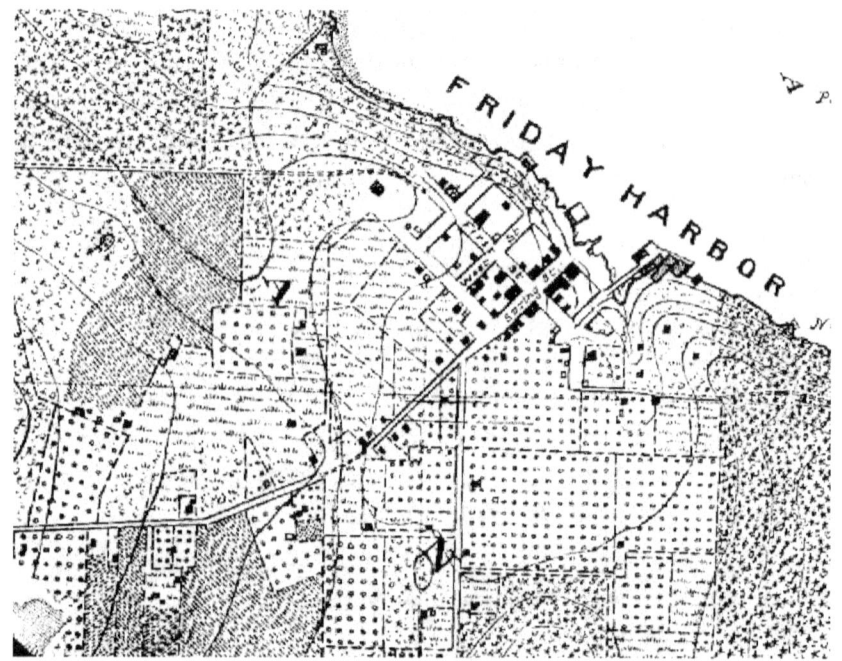

USC&GS Topographical Map of Friday Harbor Showing Orchards, 1895

Orchards in Eastsound, Orcas Island ca. 1900

dairy industry and consequent rise in beef cattle, as well as the decline in draft animals (oxen first, then horses and mules). This implies that oats were the preferred feed for dairy cows and horses and that barley was better for beef and sheep.

Fruit

Although farmers planted orchards on their homesteads as early as the 1860s and 1870s, the fruit industry in the San Juan Islands did not really take off until the 1890s. This was due to the growing market for fruit on the West Coast and, with advent of the railroad, nationwide. Production began in the mid-1800s in Oregon with stock developed and supplied by Dr. John McLoughlin at the Hudson's Bay Company Farm at Fort Vancouver. Orchards had become well established in western Oregon by the 1860s, but large-scale production in Washington Territory did not really begin until the 1880s, and, according to census figures, the annual value of fruit production nearly doubled in the 1890s.

In the San Juan Islands, Italian prune plums, which had become very popular through the efforts of the Portland horticulturalist J. R. Cardwell, were first planted on Orcas in the mid-1870s. These were followed by apples and pears, as well as apricots, cherries, peaches, and regular plums. Berry fruits such as blackberries, raspberries, and strawberries were also grown in the islands.

Details of the local industry can be gleaned from first-hand accounts offered in letters, journals, and memoirs. James Francis Tulloch's career is described in his memoir (*The Diary of James Francis Tulloch, 1875–1910*). Tulloch, the son of a Methodist minister, moved to Orcas Island in 1875 when he was 27 years old, and married and settled on a farm near Eastsound. In 1877, he planted 300 apple trees, the beginning of what would become one of the largest orchards in the islands, totaling 1,700 trees. In addition to cultivating his own fruit trees, he also hired out to prune and tend orchards for others.

> ### C. E. Cantine
>
> *When he came to the island the forest extended down to the water in a dense labyrinth of trees standing so thick on the ground that it was almost impossible for a person to penetrate it. By hard labor and indefatigable efforts he has turned the wilderness from a waste of trees to acres of fruit-bearing orchard. Here he has 1200 thrifty and productive trees embracing varieties of apples, cherries, prunes and pears, which he ships to the city markets. Last year, which was not a very good year for fruit, he shipped 300 boxes of apples, one-half ton of cherries, 1,500 pounds of prunes, one-half ton plums and forty cases of strawberries, all of which came from a six-year old orchard.*
>
> —*The San Juan Islands, Illustrated Supplement to the San Juan Islander*, 1901

Another, very detailed description of fruit farming is revealed in the correspondence between Charles Edward ("C. E.") Cantine and his son, E. J. ("Ed"), extant records of which date from 1894 to 1904. In 1893 C. E. Cantine moved to Lopez Island when he was 56 and commenced clearing the land he had purchased near Lopez Village. He then planted apple, cherry, and pear trees, forming the largest orchard on Lopez Island.

They are setting apple trees in rows 32 feet apart–running North/ South & the trees 12 ft apart in the row. I think I shall set my apples 25 ft. apart, E&W & 25 ft. N&S & then put in cherry, plum, pear etc. between the trees 28 feet apart; which would make rows 25 ft apart & trees in rows 14 ft—I let the trees crowd, the cherries etc. could be taken up. That would give us nearly double the number of trees. Where clearing is so expensive we ought to set us thick as will bear.

— C. E. Cantine to his son Ed, February 5, 1896

The first task for James Tulloch and C. E. Cantine, as well as other fruit growers, was the establishment of proper land for the trees: cutting down the existing forest, slashing brush, and pulling stumps. Once the stumps were pulled, they had to be hauled, piled, and burned. Then the cleared land was plowed, harrowed, and planted. All this was an arduous and expensive task.

The varieties of fruit planted depended on what was available, fashionable, and marketable. Orchardists often expressed their preferences in letters or opinion pieces in the newspapers. For apples, Cantine chose mostly Baldwin, but he also planted one or two rows of Russets (Roxberry or English), and one each of Northern Spy and Waxen, as well as one of Red Astrackan for summer, and one of Rambo and Gravenstein for fall. Faced with the perennial problem of getting supplies from off-island, he discussed the relative merits of several nurseries: two-year-old vs. one-year-old trees; freightage; and reliability.

EAST SOUND NURSERY

Look at these prices and compare with others. Trees are all first-class; good as the best.

Apple trees	4 yr old		15c.	$1.25	$10.00
"	3 "		12½	1.10	8.00
Prune "	2 "		15	1.25	10.00
Cherry "	2 "		25	2.25	18.00
Pear "	2 "		20	1.75	15.00
"	3 "		25	2.25	18.00
Plum "	2 "		20	1.75	15.00
Peach "	2 "		20	1.75	15.00
Quince "	2 "		25	2.25	18.00
Blackberry				50c	2.50
Red Raspberry				50c	2.50
Black Raspberry				50c	2.50
Fay's Prolific Currant				1.00	5.00
Black Cherry Currant				50	2.50
Gooseberry				50	2.50

Strawberry, 50c per 100; $4 per 1000
Rhubarb, $1.25 per dozen.

The above prices are very low for first class trees but we will discount the above prices ten per cent. on all orders amounting to $10.00 or more (accompanied by the cash) received on or before Jan. 1st, 1895.

Later, in a detailed accounting of costs, including 300 apple trees at six and a half cents each; 206 cherry, pear, and plum trees at four cents each; and 88 prunes at two cents each, he enumerates several varieties: Bartlett, Beurre D'Anjou, Beurre Easter, and Flemish Beauty pears; Peach, Bradshaw, and Columbia plums; and Tragedy, Silver, Italian, and Petite prunes. The proliferation of nurseries in the Northwest had certainly driven costs down by the time Cantine planted in the late 1880s and early 1890s. When Waldron Island homesteader Sinclair A. McDonald bought his stock from the Mitchell & Johnson Nursery in Victoria in November of 1870, he paid a little over 50 cents per tree for apples, 75 cents per tree for pears and plums, and one dollar per tree for cherries.

Fruits Good For San Juan

[E=Early; M=Middle; L=Late]

Apples—Gravenstein (M), Gromes's Golden (L), King (M), Northern Spy (L), Olympia (L), Ortley (L), Wagner (L), Yellow Belleflower (L), Yellow Newton (L), Yellow Transparent (E).

Pears—Anjou (M), Bartlett (E), Clairgeau (L), Comice (M), Flemish (M), Seckle (M), White Doyenne (M), Winter Nellis (L).

Sweet Cherries—Ring (M), Black Republican (L), Hoskins (L), Lambert (L), Royal Anne (E).

Sour Cherries and Dukes—Early Richmond (E), Montmorency (M), Northwest (L), Olivet (M), May Duke, Late Duke, Reine, Hortense.

Peaches—Alexander (E), Charlotte (E), Early Crawford (E), Triumph

Apricots—Gibb (E), Moorpark (E)

Plums—Abundance, Bradshaw, Peach, Wickson.

Prunes—Hungarian, Italian, Silver.

— San Juan Islander, March 15, 1912

Clark Farm Prune Dryer on Orcas Island

Prune Dryers

Island orchardists used fruit dryers or evaporators to dry their fruit crop—particularly prunes—for storage and shipment. Dryers were a Northwest innovation, because there was not enough sunshine to dry the prunes as they could be dried in California. Borrowing from the designs of hop dryers in Oregon, which used large, barnlike structures to waft heated air through lofts containing the hops, prune evaporators were two stories in height, with steeply pitched pyramidal roofs topped with ventilators that allowed heated air to rise through slats piled with fruit. Orcas Island had the majority of prune dryers: in 1898, the *San Juan Islander* listed eight evaporators there—Seattle Fruit Co., Peters Evaporating Co., Wm. Hambly, W. O. Clark, M. J. Reddig Co., John Bergman, Zangel Co., and Adkins & Co. Eventually, on Orcas fruit dryers were located in Crow Valley (Wm. Hambly), Doe Bay (George Culver's Alderbrook Farm), Eastsound (the Langell and Pike Farms), and West Sound (Adkins & Co., C. B. Buxton, W. O. Clark, and M. J. Reddig). On San Juan, G. B. Driggs had a fruit dryer on his land on the southern edge of Friday Harbor, and a prune dryer was built by C. M. Tucker near the Argyle Mill. George Griswold also had one on his farm on Shaw.

According to the recollections of Roy Kimple, E. L. Von Gohren, who established and maintained the orchards of Reverend S. R. S. Gray on Orcas Island, urged local growers to concentrate on a half-dozen varieties of apples, to make sorting and packing easier. He recommended Gravenstein, King, Wagner, Rhode Island Greening, and Spitzenberg, with Blue Pearmain in some locations. Another source recommended other apple varieties, depending on the harvest season: summer—Oldenburg Summer, Summer Pearmain, and William's Favorite; fall—Dutch Mignonne in addition to Gravenstein and King; Early Winter—Rhode Island Greening in addition to Yellow Belleflower and Blue Pearmain; and late keepers—Spitzenberg in addition to Northern Spy and Yellow Newton.

Orchardists in the San Juan Islands grew several varieties of plums, including Italian, Silver, and Hungarian, although the Italian was particularly favored for prune production. Prunes were harvested in early September and taken to the dryers for washing and drying. (Dryers were located near streams, springs, or wells for a fresh supply of water.) The fresh fruit was first weighed, then immersed briefly in a boiling water–and–lye solution, dipped in clear rinse water, and then transferred to a shaker, where it was

G. B. Driggs's Prune Orchard

evenly distributed on trays. After drying, the prunes were shipped off-island. Some evaporators had the capacity of seven tons per day. In 1898, the *San Juan Islander* reported that the prune crop for Orcas Island was nearly 100 tons, which, at four cents a pound, yielded $8,000—a substantial crop. Rail connections with the Midwest and East Coast accessed additional markets for the vast number of prunes being produced.

Care of planted orchards was an ongoing, time-consuming task. Both C. E. Cantine and James Tulloch mention pests of both the insect and four-legged varieties. Cantine, for instance, describes removing caterpillar nests and eggs, as well as spraying with ammonia and sulphate of copper for powdery mildew. Tulloch devotes a lot more ink to the subject. He solved the problem of tent caterpillars by applying a ring of coal tar over a mix of flour paste and soft soap (so the tar would not damage the young tree trunks); once the caterpillars were knocked down, they were unable to get up past the ring of tar. Tulloch had far more difficulties with deer and resorted to several methods of discouraging them: still-hunting, hunting with a fire jack, setting guns, and setting stakes. The first two methods involved hunting with guns—either waiting quietly near trails or dazing deer at night with an iron basket full of lit

Corner of Malcolm and Argyle in Friday Harbor

pitchwood carried on a pole over the shoulder. Setting either guns (with a trip line tied to the trigger) or stakes (sharpened and set in the ground pointed toward where the deer would jump over the fence) was far more dangerous, in that either method was as likely to hurt or kill a person as a deer.

Early orchardists also built fences to keep out the deer. "Whispering Pete" Serry, so named because of his booming voice resulting from working in a sawmill, built a ten-foot-high cedar rail snake fence around his orchard on Waldron. Early twentieth-century photographs show both Friday Harbor and Eastsound surrounded by extensive orchards, which are enclosed within tall

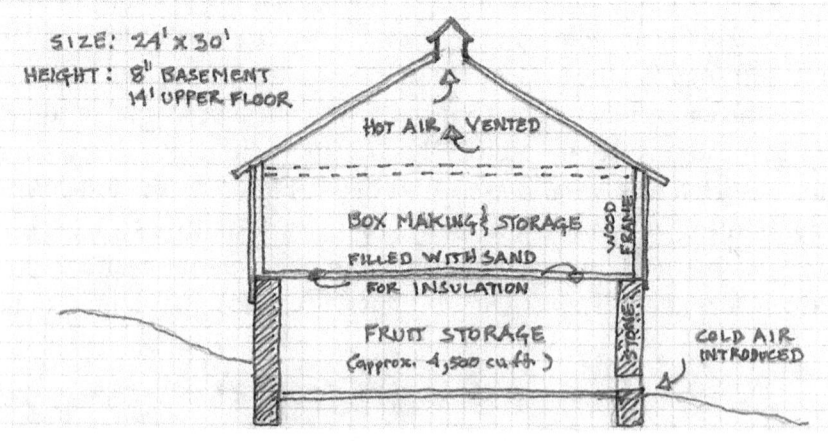

Cross Section of Hambly Apple House, Crow Valley, Orcas Island

Fruit Barns

Farmers built barns specifically designed for the storage of fruit prior to shipment. Several of these buildings are of frame construction, consisting of thick (2" x 6") walls insulated with sawdust, with 12"-square openings in the ceiling leading to ventilator shafts through the roof. This allowed warmer air to rise while retaining cooler air in the storage area. Some had a lower story of stone construction, with vents near the floor level that allowed cool air to enter and the warm air to vent above.

[George W. Meyer] received last week the returns from a shipment of 648 boxes of pears made to the Earl Fruit Company and sold in New York City for $2,607.50, averaging for the entire shipment over $4.00 a box. Of the entire lot 564 boxes were of the famous Beurre d'Anjou variety, of which Mr. Meyer makes a specialty. The selling prices of these ranged from $3.37½ for boxes containing 140 pears to $4.62½ for boxes of 70.... Freight charges at $1.50 per hundred pounds, selling agents' commission at seven percent, and cost of icing in transit amounted to $708.78, leaving Mr. Meyer a net return of $1,902.72 for the 648 boxes, or very nearly $3.00 per box or over $9.00 per tree.

San Juan Islander, November 18, 1910

(eight feet or higher) picket fences. These pickets were made from cedar that was split or riven and then sharpened at one end—an easy method prior to the proliferation of sawmills in the region.

Marketing island fruit was always a challenge because of the cost and difficulty of transporting it. While the principal means

Spraying Mrs. Waldrip's Orchard on Orcas Island

> *Mr. George W. Meyer is very busy at present planting pear trees. Two weeks ago, where he is now setting the pear trees, was to be seen one of the finest prune orchards on the island. Mr. Meyers thinks there will be more money in raising pears, so he sharpened his ax and slashed the prune trees right and left. Mr. George Gibbs is also digging out his prune trees.*
>
> — The Islander, March 19, 1896

of transportation in the region was shipping, there was more successful competition for regional markets on the mainland until the railroads arrived. Tulloch's chief market was Seattle, although he also shipped to Bellingham, Port Townsend, Tacoma, and Victoria. Always wary of the commission agents—"These fellows pay a month's rent in advance and hang out their signs and are ready to fleece the unwary farmer and fruit growers out of thousands of dollars"—he made a habit of accompanying his shipments, and relates several amusing stories of getting the better of those who were trying to take advantage of him. Tulloch estimated that he shipped a total of 75,000 boxes of apples over his 35-year career, totaling 3,000 to 4,000 boxes annually at its height in the first decade of the twentieth century. Overall annual shipments from Eastsound on Orcas Island were estimated at 25,000 to 30,000 boxes per year each of apples and pears, with additional thousands going out from Olga, Orcas Village, and West Sound.

The market for fruit varied considerably during the last decade of the nineteenth century and the first decade of the twentieth century. In 1897, while C. E. Cantine mentions that the Klondike gold rush had pushed up the price of eggs—20 cents per dozen in Seattle compared to 11–15 cents in New York City, 15 cents in San Francisco, and 11 cents in Chicago, he noted that prunes were only $10 per ton and apples and pears ranged from 25 cents to 75 cents per box, or bushel. However, by the early 1900s local growers were getting excellent prices for their fruit. In early 1907 a local paper reported that in the previous year about $100,000 worth of fruit had been shipped from Orcas Island alone: 75,000 boxes of apples (at 80 cents each); 12,000 boxes of pears (at $1.25 each); and 125 tons of prunes (at $100 per ton).

The prune industry was one of the first to succumb to the vicissitudes of regional and national economic forces. Because of the heavy planting in western Oregon and Washington in the 1880s, increased production coincided with the national depression of the mid-1890s (precipitated by the Panic of 1893). Fruit growers such as Cantine, Tulloch, and Von Gohren then moved into apple and pear production. The apple industry peaked in the San Juan Islands around 1910, when the census noted 76,731 apple trees; that same year western Washington still produced about a third of the total state crop. This apogee was marked by an event that highlighted the prosperous future of the region: the Alaska-Yukon-Pacific Exposition of 1909. Held in the fall of that year on the future University of Washington campus in Seattle, the AYP was a world's fair that drew thousands to view the economic abundance of the region. San Juan County had its fair day on July 14, and most of the county's exhibit was devoted to the local fruit industry. The *San Juan County Exhibit World Fair Visitor Register* provides eloquent testimony to the thousands who visited the exhibit from almost every state in the United States and from many foreign countries. Ironically, San Juan County brought home a

Berry Pickers near Flat Point on Lopez Island, ca. 1920

Olga Strawberry Barreling Plant

Strawberries were picked, hulled, and packed in flat crates in the field and then brought to the barreling plant in trucks. Registered and weighed on a floor scale just inside double doors on the west side of the building, the berries were poured into a tank of running water and then sprayed clean on a chain-metal conveyor belt before being graded on a rack. Women hand-sorted out the green and nonhulled fruit, and then another belt sorted the berries into small, medium, and large. The fruit fell from a trough into continuously jolted barrels, packing them solidly; each 425-pound barrel had 318 pounds of berries layered alternately with 107 pounds of sugar. The barrels—about 17 per day—were loaded from the east dock onto trucks, shipped by ferry to the mainland, and driven to Everett for freezing.

banner, which still hangs in the courthouse, that reads Best Cherries at the Expo—even though cherries were not as major a crop as apples and pears.

There was a brief period of berry farming on Lopez Island during the 1920s. After the Salmon Banks Cannery at Richardson burned down in 1922, Ira Lundy planted loganberries on the fields west of the dock and store. This offered the out-of-work cannery hands employment picking fruit. According to Marguerite McCauley Goodrow, who first picked berries there when she was seven years old, she was paid 18 cents per flat crate; each crate held 12 boxes. Apparently, she was a good enough picker to earn $25 her first summer and she continued picking every summer for many years.

But the irrigated, better soils east of the Cascades produced apples, cherries, and pears with more color (although inferior in taste, according to island farmers), and by the 1930s the islands were only producing a third of what they had achieved 20 years earlier, while western Washington as a whole only contributed 2 percent of the total state crop. According to the federal census of agriculture, county pear production peaked at 10,070 pear trees in 1925, and then steadily declined for the next two decades. By 1950, there were slightly more than 5,000 apple trees—less than today—and fewer than 2,000 pear trees.

An important fruit crop during the 1930s, 1940s, and 1950s was strawberries. In the early 1930s, farmers in the Olga area of Orcas Island began growing Marshall strawberries, eventually bringing some 450 acres into cultivation. Growers produced both plants and berries. In 1936, W. P. McCaffray of the National Fruit Canning Company approached the Orcas Island Berry Growers Association (est. 1936), proposing to supply the building materials and machinery for a barreling plant on land to be purchased (by Glen Rodenberger) and provided through the association. Construction of the Strawberry Barreling Plant began in 1937 and was completed in 1938, the year rural electrification came to Orcas Island through the Orcas Power & Light Cooperative. The facility provided work for scores of Orcas residents who cultivated, harvested, processed, and hauled local strawberries. Operations lasted until the end of World War II, when a combination of disease in the strawberry plants and shortage of wartime labor led to the plant's closing. A brief period of strawberry farming occurred towards the end of World War II on San Juan Island, where George P. Jeffers, owner of the San Juan Canning Company in Friday Harbor, bought 710 acres of San Juan Valley farms and planted 40 acres in strawberries. Unfortunately, this operation only lasted a year or two due to unfavorable market conditions.

Dairying in San Juan County:
Rapid Development during the Past Three Years

The rapid development of dairying in this county, and especially on San Juan island, during the past three years is a source of a great deal of satisfaction to the Islander, which had labored long and patiently to secure the establishment of a creamery here and to persuade farmers to engage in this lucrative branch of agricultural industry and build up their farms by stock raising instead of exhausting their soil by grain production year after year. The Islander began preaching the gospel of dairying [when] not a can of cream was being sold in the county or shipped from it. Now nearly all of the fine farms of San Juan, Orcas and Lopez islands are largely devoted to dairying and stock-raising and cans of cream constitute a considerable part of the freight handled by the regular steamers at nearly all points touched. On this island alone there are 600 or more milch cows and the number will probably reach 1,000 within the next two or three years. On Lopez island, which is next in importance in milk production, there are several hundred cows, furnishing a large supply of cream to an excellent creamery at Lopez, the proprietors of which, Messrs. Bugge & Frederickson, have secured an option upon a site for the establishment of a creamery here. Large quantities of cream are shipped from here to Anacortes and Seattle, and John B. Agen, of Seattle, one of the wealthiest and most extensive creamery operators in the state, has had a representative here this week arranging to test and purchase cream here for shipment to Seattle until he can build and operate a first-class creamery here.

— *San Juan Islander*, April 21, 1906

Dairy

Although practically every homestead had a milch cow or two, commercial dairying soon became a significant component in the San Juan Islands' agricultural economy. The number of dairy cattle countywide gradually rose from 247 in 1880 to 443 in 1890 and 774 in 1900, jumped to 1,916 in 1910, and peaked at 3,175 in 1920. Increased dairy production in turn led to the establishment of a creamery in Friday Harbor in 1901. Three years later, the local newspaper reported that the creamery had converted 3,000 pounds of cream to 1,000 pounds of butter in a single day. It was so busy it had to introduce a night shift so the creamery could run 24 hours a day. By 1907, the creamery produced 120 tons (240,000 pounds) of butter annually. Other island creameries were established in Eastsound (1901), Shaw (date unknown), and Lopez (1903).

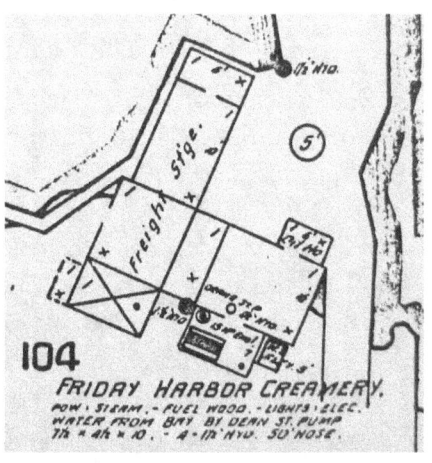

1907 Sanborn Fire Insurance Map

Growth in the dairy industry led to the need and desire for improvement in stock and methods. In an October 8, 1910, meeting attended by State Dairy Inspector L. W. Hansen, 19 dairy farmers formed the San Juan County Dairymen's Association. With the introduction of the first Washington State College (now Washington State University) County Extension Agent in 1921, there was a big

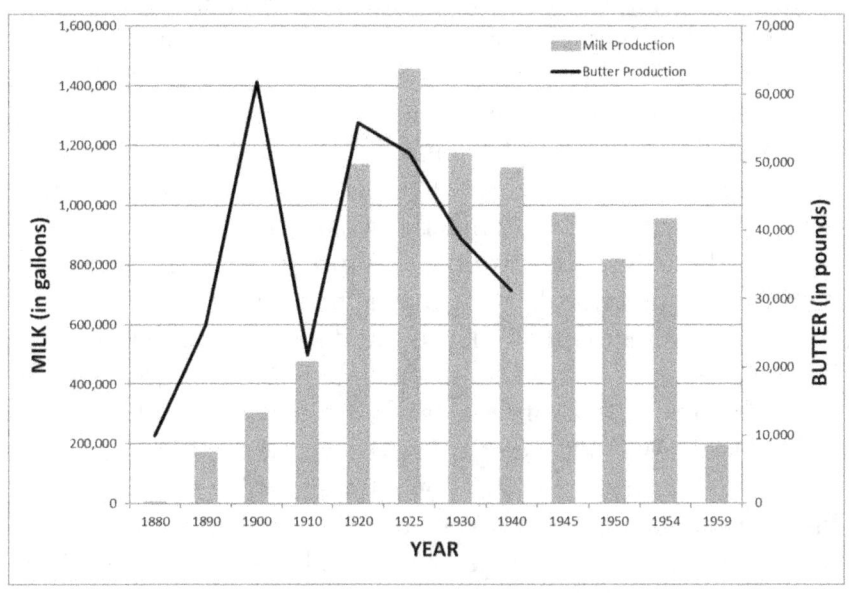

Milk Production (in gallons) and Butter Production (in pounds) in San Juan County, 1880-1959

> ### *The Lopez Creamery*
>
> *During the past few years the ranchers of San Juan County have more clearly recognized than was previously the case the scope and financial advantage of dairying and many farmers are to be found who at one time kept a solitary cow for their own household purposes who now derive a considerable portion of their income through the creameries which do business in this territory. The success of the Lopez Creamery well illustrates the change in point of view and practice for the record achieved since its establishment three years ago, would have been impossible in earlier days. The company of which Frederickson & McKinnis are the proprietors is now sending nearly 1,500 pounds of butter to the markets of Seattle, Bellingham and Anacortes each month and experiencing a steadily increasing demand for its brand.*
>
> — *Islands of San Juan County, Washington, Everett Morning Tribune,* 1906

push to improve stock through selective breeding, animal health through tuberculosis testing, and production through cream content testing. The San Juan County Jersey Club was organized in 1922, and in August of the same year the San Juan Islands hosted the Pacific Northwest Guernsey Breeders Association. The reformation of the San Juan County Dairymen's Association in 1924 helped this movement by providing a forum for farmers to discuss improvements to their stock, animal health, and production.

Dairy farms in the San Juan Islands had relatively small herds, ranging from 6 to 12 milch cows. Aside from data provided through the federal census of agriculture, some anecdotal information helps describe dairying during this period. In 1920, for instance, the average daily milk produced per cow was 543 pounds. James Cousins on Lopez Island, a member of the San Juan County Cow-Testing Circle, kept records of his dairy herd from June 1922 to June 1923. With an average of nine cows, he got 4,908 pounds of milk that yielded 239.2 pounds of butterfat, which earned $108.76. Figuring in $66.01 of cost in "succulent roughage" (apples), dry roughage, oats, and pasturage, this yielded a net profit of $42.75 or $4.75 per cow. L. L. Salisbury, of Friday Harbor, entered a month-long milk contest in 1936: with his three Guernseys he averaged

Dairying In San Juan County

Throughout the recent period of financial unrest, while Wall street brokers daily dreaded the opening of market and bankers groaned under the burden of their responsibilities, the dairy men of San Juan county lived in serene disregard of outside conditions and enjoyed uninterrupted prosperity in the conduct of their chosen business.

— Islands of San Juan County, Washington, Everett Morning Tribune, 1906

1,212 pounds of milk and 52.8 pounds of butterfat each. Altogether, from 1920 to 1940, yearly milk production in the islands averaged over one million gallons.

State regulations regarding dairy production led to changes in farmstead layout and building design. In 1895, the Law Relating to Dairy Products was passed, which mandated the labeling and dating of butter and established the office of the State Food and

Pacific Northwest Guernsey Breeders Association
1922 Tour of the San Juan Islands

Milk Houses

Milk houses were designed to provide a clean, cool place apart from the milking shed or parlor. Fresh milk comes out of a cow at around 90°F; however, in order to prevent the growth of bacteria that spoils the milk, it must be cooled to below 50°F. In order to do this, milk was carried in buckets or cans to the milk house, where it was placed in a trough of cold water. Early on, milk was taken to a creamery, where the cream was separated from the whole milk. Around 1900, individual machine separators, usually centrifugal, were offered at affordable prices to farmers. At first these were hand operated; later they were driven by gasoline engines or electrical motors.

Milk houses were both separate from the barn—usually on the side away from the manure removal area and near a farm lane for convenience of cream pick-up—and close enough so that the milk did not have to be carried too far. Sometimes the milk house was built right next to the barn but separated by a passageway or porch. Ranging from 7 to 12 feet in width and 7 to 16 feet in length, most were 8' x 10' and constructed on concrete stem walls ranging from 1 to 3 feet in height. The common construction is 2x4 stud walls with either shiplap or clapboard (and sometimes shingle) siding; gable roofs were first covered with wood shingles and later metal. (Some exceptional milk houses were constructed of concrete block.) On the front, there was sometimes a roof hood or porch extending over the entrance and a poured concrete pad to set the cans on for pickup. Some milk houses featured a raised platform for mechanical separators and sinks for washing containers. The law required whitewashing on the interior, a screened door, and screened operable windows.

Dairy Commissioner. Following World War I, major legislation was passed in 1919 to regulate dairy farm layout: it mandated clean water, sanitary equipment, whitewashed milking areas, and a milk house separated by at least 50 feet from any filth. This led to barn designs with clean, whitewashed stalls and stanchions with gutters for flushing out manure and urine. Dairy farmers built stand-alone

> In the conduct of dairy work special conditions of soil are necessary. One must have a ranch with moist, arable land, level enough for comfortable tillage, upon which to grow an abundance of necessary feed. The land must be plentifully supplied with water, flowing water at all times to be preferred, and if in addition to these things, the ranch contains some hillside land for an early run, then the conditions are ideal. Many such favored spots are to be found among the islands of San Juan county. The famous valley on San Juan island is an especially good tract of several thousand acres. Lopez island also has a large acreage which now is primarily devoted to dairying; perhaps today she leads in the number of dairy cows and the value of dairy products. Orcas, with her more broken surface, is principally devoted to fruit growing, but here also many fine dairies are to be found, particularly in Crow valley and their number steadily increases.
> — *Islands of San Juan County, Washington, Everett Morning Tribune,* 1906

milk houses for separators, and, after the introduction of concrete, they built troughs to hold cold water in which the 10-gallon milk cans were kept cool.

Technological changes also affected dairying in the San Juan Islands. The introduction around 1900 of the Babcock Tester, which measured the butterfat content of the milk from each cow, enabled dairymen to selectively improve their stock by breeding and culling. Around the same time, Gustaf de Laval invented the centrifugal hand separator, which helped individual farmers separate cream on their farms. One of the earliest separators was bought in 1899 by Orcas islander Robert Caines, who made butter for the crews at his lime kilns. Soon the newspapers were full of ads for De Laval, Meining, and Empire separators, with capacities of up to 700 pounds per hour. Caines's fellow Orcas farmer Estyn M. Chalmers also bought a separator and con-

```
       BUY YOUR
   Milk and Cream
         FROM THE
    SWEET PEA DAIRY
       and be assured of
     Pure and clean milk
  Telephone 154    J. J. SANDWITH
             FRIDAY HARBOR
```

Howard Shull Operates a Model Dairy Farm: Modern Methods and Equipment Lessen Farm Work

During the past week the writer had the pleasure of visiting the Howard Shull dairy farm and was amazed at the ease with which a large herd of dairy cattle can be attended to by up-to-date methods, aided with modern equipment.

Mr. Shull's barn is built along modern lines with a view of saving unnecessary steps and yet sacrificing nothing for space. He uses a power milking machine and at present is milking from 15 to 20 cows in less time than three men could do the work by the old hand method.

The separator room is in the pump house. Here besides having plenty of cold water, he also can have boiling water, heated by the engine while running the milking machine. From the milk house he has a water pipe running to the hog barn where all the separated milk is run, likewise drinking water for the hogs. This alone is a labor saving means, as Mr. Shull endeavors to put on the market several pounds of pork in the course of a year.

At the hog barn he has an ingenious home-made float valve arranged so that when the milk and water receptacles are filled, the flow will automatically stop, thus eliminating waste of milk or a chance of running the water tank dry.

At present arrangements are being made to install a modern acetylene gas lighting system, which will light the residence as well as several of the other farm buildings where light is needed.

— Friday Harbor Journal, September 28, 1922

tracted with Andrew Newhall, owner of the steamer *Islander*, to ship his cream to Whatcom (now Bellingham), where Chalmers persuaded the local wharfinger at the Sehome dock to haul it to the Whatcom creamery for processing. Steam-powered separators soon followed.

Dairying was intense farming because of the amount of time and work it required. For smaller operations of six to seven cows—the average for most island dairy farms, women did most of the feeding and milking and cleaning the separator equipment. For operations of 20 animals or more, everyone in the household leant a hand, including children. It has been estimated that dairy farmers moved approximately 20 tons of milk, feed, bedding, and manure per cow per year, requiring some 60–150 person-hours, depending on the efficiency of the operation. The more efficient the design of the dairy farm, the less labor was expended on the day-to-day operations.

The center of dairy operations was the barn. In general, the desirable attributes of a dairy barn included siting on firm, well-drained ground; proximity to other structures (such as the milk house), yet distant enough in case of fire; convenient access to yards and pasture; and ample feed and bedding storage. Some handbooks advised that the lengthwise orientation of a dairy barn be north-south, to afford maximum exposure to sunlight, as well as "bank barns," which were built into sloping ground to take ad-

---

Another Purebred Bull for San Juan County: Woodworth Brothers Purchase a Fine Animal

The Channel View Jersey Farm of Orcas, announce the sale to Woodworth Bros. of West Sound, of the young purebred Jersey bull, "Brilliant Spray's Pogis," No. 201,615. This animal has exceptional breeding, his dam, "Imported Brilliant Spray," being without question one of the greatest show cows in the west, and in addition has a register of merit record of 546 pounds of fat. She was imported from Lord Rothschild's great herd of Jerseys in England, and has won many prizes on the Pacific coast in the six years she has been shown.

Woodworth Brothers are to be congratulated on having secured such a well bred bull, backed by high production to head their dairy herd.

— Friday Harbor Journal, February 23, 1922

Modern Dairy Barns

The early twentieth century introduced modern, or "scientific," designs for dairy barns, which featured a lower floor for the animals and their stanchions and a tall upper loft for the hay. The optimum width was determined to be around 36 feet, which allowed for two rows of stalls with alleys, illuminated by light on both sides, on the ground floor. The full second story was occupied by the hayloft, expanded because of the greater volume provided by a gambrel, Gothic, or bow-shaped roof. Both natural lighting and ventilation were important factors before electricity became widespread. Some experts advised four to six square feet of window per cow; others simply suggested one window for each animal. Voiding odors was accomplished by means of ventilators on the roof and in some cases shafts that conducted the foul air from the lower floor to the vents.

With the advent of "scientific" farm management systems, barns were modified in response to innovative dairy farming methods. Dairy farmers were encouraged to use concrete floors and metal pole stanchions and dividers for milking areas because they were easily cleaned. Land-grant college agricultural experiment stations, cooperative extension offices, and farm equipment manufacturers all offered specific guidelines for the layout of stalls and stanchions. For instance, it was generally recommended that the animals be oriented "head in"—towards the center of the barn—with an average stall being 3'-4" in width and 9' long, the latter divided into a 2'-3" manger, a 4"-high stall curb, a 4'-8" stall platform, and a 1'-4" gutter, or manure trough. The floor level stepped down between the manger and the platform and the platform and the walkway beyond the manure gutter.

vantage of the difference in elevation to access different levels of the structure for different functions.

Dairy cattle were brought into barns twice a day for milking. To accommodate this, sheds, usually 15 to 16 feet wide, were built abutting the haymows, with stalls and stanchions lined up in a row. A typical stall section included an alley or walkway for carrying feed to the mangers, the manger itself, a stanchion for holding the milch cow in place while it was being milked, a platform for the animal to stand on, and then another alley or walkway with a gutter or trough for the manure that the cow dropped while being milked. The material for these elements was predominantly wood. The stanchions often consisted of an upright 2x4 that was fixed top and bottom to the rails, with a pivoting 2x4 pinned to the bottom rail 7" or so away; when the cow's head entered the space between the boards, the moving member was pivoted toward the fixed member and held in place by a notched swivel on top, thus securing the animal's head in place. Platforms were constructed of thick (two inches or more) planks to hold up to hoof wear. As required by law, all wall and ceiling surfaces were whitewashed with a lime solution for cleanliness.

Ruth Guard Milking

Dairy Cow Milking Stall Recommendations

Breed	Width of Stall	Length of Platform		
		Small	Medium	Large
Holstein	3' 6" to 4' 0"	4' 8"	5' 0"	5' 6"
Shorthorn	3' 6" to 4' 0"	4' 8"	5' 0"	5' 6"
Ayrshire	3' 6" to 3' 10"	4' 6"	5' 0"	5' 4"
Guernsey	3' 6" to 3' 10"	4' 6"	4' 10"	5' 2"
Jersey	3' 4" to 3' 6"	4' 4"	4' 6"	4' 8"
Heifers	3' 0" to 3' 2"	3' 8"	4' 0"	4' 2"

N. S. Fish, *Building the Dairy Barn*
Wisconsin Agricultural Experiment Station Bulletin 369, 1924

Silos

Silos are a relatively recent farm building. Experimentation in Europe in the 1870s with the preservation of green fodder—called "ensilage," or simply "silage"—led to its adoption in the US, first in the Midwest and then later in the New England and New York regions. By keeping the fermented contents (silage) in a low-oxygen environment, farmers prevented mold and decay and preserved the nutritional value of the fodder.

Silos varied a lot in form and construction. Early versions in the Midwest and New England and New York regions were lined pits. However, farmers soon learned the advantages of vertical, or "tower," silos—the downward pressure of the silage compacted itself, thus leading to less air pockets that produced spoilage. Changes in design reinforced these advantages: tower silos were originally rectangular, but round models—originally with horizontal bands, later with vertical staves like a water tank—eliminated corner pockets of air and withstood the outward pressure of the weight of the silage. When they were placed on concrete foundations with floors, another problem was eliminated: rotting wooden foundations. Round tower silos were also constructed of concrete or structural tile blocks. These ranged from 12' to 14' in diameter and rose 35' to 40' in height; it was estimated that a 12' x 38' or a 14' x 30' silo would serve about 25 cows.

In 1905, the S.T. Playford Company of Elgin, Illinois, developed silos constructed of cement (technically concrete) staves. These are the most common surviving silos in the islands. The staves, 10" wide by 30" tall and 2½" thick with a V-shaped mortise and tenons on alternating edges, interlock to form a uniform cylindrical tower, commonly 10' in diameter. The staves were alternately overlapped 2 inches vertically, and 3"-wide steel hoops or ½" metal rods that covered these ensured a tight, integral structure. The inside was plastered with a thin cement stucco or wash to create a smooth, airtight surface. These silos were approximately 12 staves, or 30', in height, and were capped by a frame gable or hipped roof covered with shingles. Farmers had to climb up a ladder on the side of the tower in order to empty out the silage from the top down; a series of air-tight 20" x 30" doors were placed in the openings down one "side" of the silo.

Harvestore

In 1945, the A. O. Smith Company developed the "Harvestore," which featured fiberglass bonded on the inside to the metal container. The Harvestore was 20' in diameter and 61' high, painted a distinctive blue, and featured a system of unloading from the bottom of the silo. In addition to their better performance, however, they were more expensive: in the 1960s, for instance, a Harvestore cost $11,302, twice the $5,435 for a concrete stave silo.

Vacuum systems were developed for milking in the early 1900s; these consisted of pumps, pipelines running along the top of the stanchions, and spigots for attaching the hoses to the milkers. The De Laval Company (named after Gustaf de Laval) made systems for generating power for lighting and running vacuum pumps for milking. Another innovation was the manure trolley, the bucket that hung from an overhead track running along the back of the stalls and out into the barnyard, where the manure could be dumped into piles or carts for hauling elsewhere. Silos were introduced to the islands as a means of keeping cows "fresh" by feeding them silage, or fermented fodder, which is nutritionally richer than hay, throughout the winter.

Joe Groll finished, the last of the week, the construction of a new cow barn at the farm home of Mrs. H.C. Smith of Kanaka Bay vicinity. The building is 40 by 70 feet and is arranged with patented stalls, which enables a person to care for the cows in every particular, without one interfering with the other.

— *Friday Harbor Journal,* January 4, 1912

Poultry

Most farmers had poultry as part of their farmsteads: a few dozen chickens and perhaps some geese and turkeys. However, during the late 1800s island farmers began to raise chickens, principally for eggs, on a larger scale. Several farms had huge flocks and specialized in exotic breeds: Ben Lichtenberg's GEM Farm raised Plymouth Rock cockerels and White Holland turkeys, and John S. McMillin's Bellevue Poultry Farm, near Roche Harbor, specialized in Crystal White Orpingtons (first bred in England in 1886) as well as geese and turkeys.

Specific structures, called poultry houses, hen or laying houses, or chicken coops, were built to house the birds. A good building was dry, clean, well ventilated, roomy, warm during cold weather, and well lit. Because hens are often moody birds, sensitive to the amount of daylight, particularly regarding egg laying, it became important to provide artificial lighting in order to encourage laying throughout the darker winter months. Poultry houses were usually located nearest the house—because farm wives often took care of the birds—and also sometimes near the garden or orchard, where the birds could be let loose to hunt for insects. Most often one story and constructed of dimensional lumber and wood siding, these structures had either a gable or shed roof. Sometimes these long, low buildings, oriented along an east-west axis, had a "saltbox" roof, where the higher façade faced south with large

Bellevue Poultry Farm, Roche Harbor, San Juan Island

windows and the lower façade faced north with smaller windows, or louvers for ventilation. One-foot-square "chicken doors" opened onto slanted boards with narrow strips of wood nailed on so the birds could walk down to a foraging area or run, usually located on the south side. Doors for farmers were located on the sides of the building. Plan dimensions depended upon the size of the flock; building manuals recommended 3–4 square feet of space per bird, so buildings ranged from 12' x 14' for a smaller flock of 3 dozen or so to 16' x 32' for 12 dozen birds. Narrow, long buildings resulted from the addition of a series of pens or cribs in a row. Inside, poultry houses featured several specialized spaces: a roosting area, where parallel strips of wood or rods offered a space for the birds to roost at night; rows of nest boxes; feeding and watering areas; and sometimes bins for dusting. In order to ventilate the houses of the strong odors and ammonia built up from manure, centrally located shafts 6–12 inches square ran from a few feet above the floor through the ceiling and out a "chimney."

> Our SPECIAL POULTRY SALE will soon be over.
>
> Those who bought in December are pleased and fortunate. We still have plenty of THOROUGHBRED BREEDING FOWLS for those who come late.
>
> COCKS   COCKERELS   TOMS   DRAKES
>
> Don't fail to secure the very best ACCLIMATED STOCK right at home.
>
> WHITE ORPINGTONS, WHITE ROCKS, WHITE LEGHORNS
> BROWN LEGHORNS, MAMMOTH BRONZE TURKEYS
> WHITE PEKIN DUCKS
> BEST IN QUALITY,   LOW IN PRICE.
> EGGS FOR HATCHING.   BABY CHICKS.
> Book your orders NOW.
> Our customers are winning the prizes at the Poultry Shows with our BELLEVUE FARM STOCK.
> Write for 1912 Price List.
>
> **Roche Harbor Lime Company**
> Roche Harbor,   -   Washington

> We grow these birds primarily for breeding purposes, but of course sell our surplus upon the market, after using whatever is necessary for our hotel requirements. When properly and carefully handled, I think turkeys can be grown here very successfully and on a very profitable basis. It requires a great deal of patience and care when the birds are very small. After they are six or eight weeks old, however, they are very hardy and easily handled. If suitable range is provided for them, they get most of their feed outside and can be grown at much less expense for feed than larger varieties of chickens. The market for turkeys is nearly always good and particularly about the holiday season. Prices are higher than for any other fowls that are grown.
>
> — John S. McMillin, Bellevue Poultry Farm
> *Dairy, Poultry and Stock Raising in the State of Washington*, 1916

Poultry Barns

Poultry "barns" came in two forms. One was a two-story, gable-roofed structure, with the lower floor used for coops, nest boxes, and roosts, while the upper floor was either empty (and used for ventilation purposes) or used to store either grain or grit, or both. An example of this type of barn from a farm in San Juan Valley measures 18' x 30'.

The other form of a poultry barn has only one remaining example in the islands, on Lopez. Constructed for turkey production, it is a two-story structure with a monitor roof. The coops are on either side of a central aisle and have wire netting. The monitor walls (clerestory) are louvered, so that the warm air and odors can rise through the center section to be vented. In the center of building, at right angles to the center aisle, is a tightly sealed granary at the second-floor level. It is accessed by a gable dormer on the side, which allows for loading feed by means of hoist from trucks parked along the side of the building.

In the 1930 census of agriculture, San Juan County reported over 36,613 chickens. In 1936, 90 percent of these were on Shaw Island. The poultry industry began on Shaw when Elihu B. and Oscar Fowler homesteaded there in 1889, and the Fowler family soon became known for raising chickens, both for broilers and egg production. A 1944 article in *WASHCOEGG* by San Juan County Extension Agent W. J. Wylie, entitled "Poultry Pioneers in Island Hopping," reported that of the 30 families living on Shaw, 18 were actively farming, and of these, 10 were poultrymen. In 1940, Shaw shipped out 3,400 cases of eggs (a case contains 30 dozen, or 360, eggs, so that meant 1,224,000 eggs total). In the year of the article, Shaw poultry farms had 8,500 layers and 8,000 fryers.

Turkeys were also a significant commercial crop: one week in 1907 George Heidenrich of San Juan Island shipped over 700 pounds of dressed turkeys to Seattle; later, in 1912, about a ton and a half of birds were shipped from the San Juan Agricultural Company's dock in Friday Harbor in a single week. In the 1930s, turkeys became a major crop, with 10,767 recorded in the 1930 agricultural census, a number which grew to roughly 20,000 in 1936. That year, the San Juan County Turkey Growers Council met with County Agent Charles T. Meenach. Alfred Douglas, president of the council, alone had some 1,500 birds on his farm. Some farmers grew a special beardless variety of barley, which yielded one ton per acre ("with rabbits") for turkey feed. While major turkey production proved ephemeral—there was an exceptional "spikelet" in the 1940s and 1950s—chicken numbers fluctuated depending on the costs of feed and transportation, but remained in the 20,000 to 25,000 range until the 1960s, when they began to fall off precipitously.

Beck Children Feeding Turkeys, Orcas Island, 1920

Lost at Bridge, Win with Peas
Joe Rubens Tells How Cannery Was Established

In 1923 there was a group that played bridge at the City club that included among others, Frank Post, Tony Russell, Frederick Wilson and John Henry. Henry won all the time, much to the disgust of the other players, and when he suggested the building of a pea cannery at Friday Harbor on San Juan Island, he had little difficulty securing financing support of his bridge opponents, including Wakefield & Witherspoon.

... Mr. Henry had raised seed peas in many sections of the west and found that the island produced an exceptional sweet and tender pea.

Mr. Henry and I were delegated to contact the farmers for acreage and were promised sufficient acreage to warrant the building of a cannery, which was built.

All finances were subscribed by the group when Mr. Henry consented to the appointment as president and general manager of the company with headquarters in Friday Harbor. The bridge players went into it with the belief that if the cannery paid no other dividends it would make them money in keeping Mr. Henry out of town and away from their bridge table.

The first pack of the first pea cannery in the state in 1923 was only 23,000 cases. But last year [1935] another cannery was built in Mount Vernon, and the pack of salt air peas this year is expected to reach 2,000,000 cases, with the product sold all over the United States.

All because a group of City club bridge players found the only way they could beat Mr. Henry at bridge was by getting him out of town.

— *Spokesman Review* quoted in the
Friday Harbor Journal, August 27, 1936

Peas

Dry peas had long been a local farm staple, and the market for split peas during periods like World War I bolstered production in the islands. Census figures from 1880 and 1890 report 1,656 and 1,316 island acres planted in peas, and in 1912, the *Friday Harbor Journal* reported that W. H. Cadwell of Lopez Island harvested 69 sacks of peas—nearly 5 tons—from 4 acres. The Argyle Mill on San Juan Island, originally developed for milling grain, even adapted its operations for pea production.

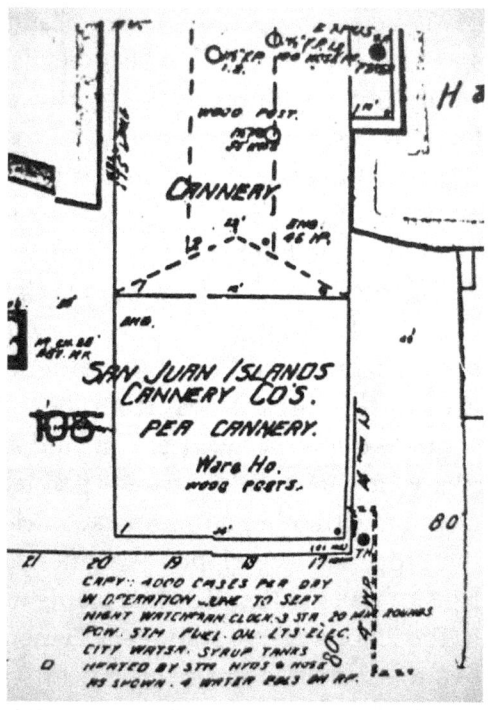

1930 Sanborn Fire Insurance Map Plan

Dealers from C. C. Morse & Co. of San Francisco, who had contracted with local farmers for the product

John Sundstrom Working Pea Viner, Gorman Farm, San Juan Valley

of 317 acres, shipped some seed peas (of the table variety); upon inspection, they were full of pea weevils, which the company claimed would not germinate in the Northwest climate, but farmers were loath to risk it.

In 1922, seeds were introduced by John M. "Pea" Henry of Spokane to San Juan Valley for the production of green peas for canning. The following year Henry established a cannery in Friday Harbor—the San Juan Islands Cannery—with the brand name "Saltair Peas." Henry's company supplied farmers with both seed peas (Surprise and Perfection varieties) and the "viners" that were used to separate the peas from the vines and pods. Farmers grew about two tons of peas per acre and sold their crop at $60 per ton. Many San Juan Valley farmers planted peas, usually in a four-year rotation with other crops for each field. Peas were not only a lucrative crop, but the stripped pea vines could be used as fodder and silage for dairy cattle. The pea plants were harvested and then taken to the 20 or so large stationary "viners," machines fed by a crew of five to six men, and then the peas were hauled to the cannery in town. Peas were also grown in Crow Valley on Orcas Island.

During the 1920s and 1930s the cannery bought up several large farms in San Juan Valley from the families of the original homesteaders. In 1939, the floor of the cannery building collapsed, sending some 10,000 cases of canned peas onto the bottom of the harbor. While these were fished out and repackaged—aided, according to island lore, by local schoolchildren—the invasion of weevils—particularly the pea leaf and the Sitona—the following year (1940) soon led to crop failure and the demise of the first phase of the pea industry. The cannery building was subsequently demolished during World War II.

Miscellaneous Crops

Island farmers experimented with several crops that did not reach the extent and production of such staples as hay and grains, sheep, fruit, dairy, and peas. These included common standards such as potatoes, tobacco, and hops as well as more exotic crops such as rhubarb, tulips, rabbits, ginseng, and seed crops.

Potatoes. Potatoes have been blamed as the cause of the Pig War. Charles Griffin at Belle Vue Sheep Farm records getting his potato seed from the Cowichans on Vancouver Island, while Lyman Cutlar at his farmstead on First Prairie claimed that he got his "American" potatoes from Port Townsend. Potatoes were a staple for homesteaders but were also raised for market. In 1898 M. R. Lundblad of Argyle on San Juan Island claimed that he raised from one potato a crop of 34, with a total weight of 22 pounds 1½ ounces. The variety most commonly referred to is Russet Burbank; Joe La Chapelle raised Beauty Hebron at the former Belle Vue Sheep Farm site. The *Friday Harbor Journal* of May 2, 1912, reported that William Beigin planted a 12-acre patch of Burbank, Mortgage Lifter, and Nettie Gem—the latter imported from North Yakima.

> ***The Spud Crop***
>
> *The potato crop on Orcas and the other islands in that vicinity is unusually good this season, and will average five tons to the acre. Mr. Robinson, of Orcas Island, informs us that from one-eighth of an acre he raised this year 1600 pounds of as fine tubers as a Land Leaguer ever worked his jaws over. Potatoes are worth about 90 cents a bushel in San Francisco.*
>
> *— Northwest Enterprise,* October 28, 1882

Tobacco. Surprisingly, several island farmers tried growing tobacco, which is generally considered a warmer-climate crop. James Fleming in San Juan Valley, Peter Bostian on Orcas Island, and two families on Shaw Island raised the weed, and Bostian even got a silver medal for his leaf tobacco at the 1909 Alaska-Yukon-Pacific Exposition in Seattle.

Hops. Hops are principally used for brewing beer, and in the 1890s there were several large hop operations in Oregon and the Puget Sound region. Farmers also tried growing hops in the San Juan Islands; the 1890 federal census of agriculture reported 14 acres in production, yielding 18,200 pounds of hops. Two farmers in San Juan Valley raised hops, one of them being E. P. Bailer near

Bailer Hill, who built a hop dryer, while the other farmer, C. H. Sutton, nearby, shared its use.

> C. H. Sutton finished spraying his hop yard on the 19th. He found a few lice and thought he would try the ounce of prevention and see if it would be worth a pound of cure, and he is satisfied that if you must spray, do it as early as possible, as the young lice are much easier killed than when they get wings. He used quassia chips and whale oil soap, giving them a good strong dose, and it is now hard to find a live louse in the field. The hops look well and he expects to harvest at least as much as last year.
> — *The Islander, July 26, 1894*

Tulips. While the Skagit Valley comes alive every spring with acres of blooming bulbs celebrating the annual Skagit Valley Tulip Festival, tulip growing in the Northwest originated on Orcas Island. This was largely due to the efforts of George William Gibbs, who was born in Tewkesbury, England, and immigrated to the United States at the age of 17, first to New York State, and then to Ann Arbor, Michigan, where he bought his first farm. Around 1886 or 1887, already well known for his widely published articles on farming in several states and territories, Gibbs chose Orcas Island

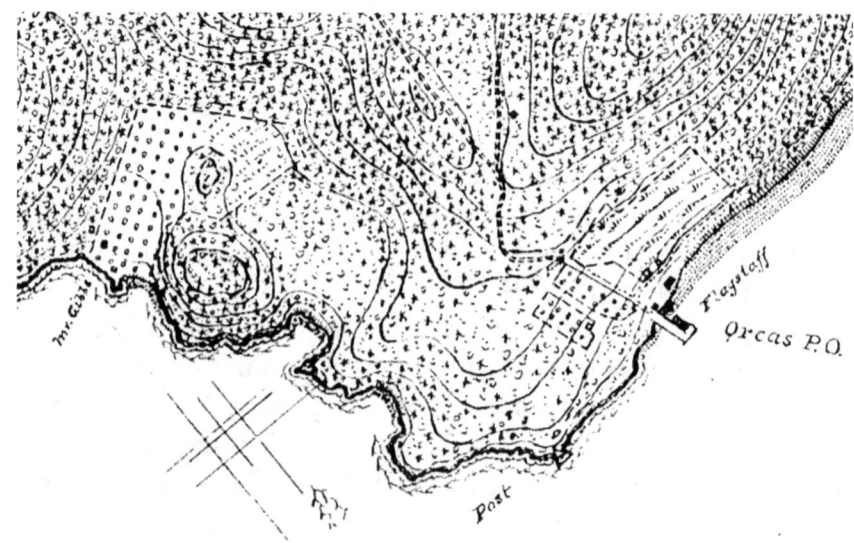

USCG&S Map of George Gibbs Farm, 1895

for his next farm. At first, Gibbs pursued fruit and nut growing, leasing 121 acres in Warm Valley near Orcas Landing from the San Juan County Commissioners. However, after planting hyacinths and noting their proliferation, in 1892 he ordered five dollars worth of hyacinth, tulip, narcissus, crocus, and lily bulbs. By 1898, he had attracted the attention of some Dutch bulb growers, who came to inspect his acreage on Orcas and were favorably impressed. Upon submitting bulbs to the Trans Mississippi and International Exposition of that year in Omaha, Nebraska, he received a silver medal for his varieties of narcissus, iris, hyacinth, crocus, and tulip, with commendations on the particularly fine quality of his Madonna lily.

In 1899, Gibbs moved to Bellingham, eventually settling in 1902 in Clearbrook, near Lynden. He persuaded the US Department of Agriculture to send him 15,000 Dutch bulbs for experi-

> Geo. Gibbs, of Orcas, is at this writing planting 50,000 Holland bulbs. This industry, by the real pluck of Mr. Gibbs, is making a stir all over the country. He has received the silver medal from the Omaha Exposition and it is an honor to its owner. Call and see it, he is very willing to show it and is proud that he is its possessor.
>
> — *San Juan Islander*, October 27, 1898

mentation in 1905, and three years later was instrumental in convincing it to establish a 10-acre test garden near Bellingham. This proved to be the main source of bulbs for the Bellingham Tulip Festival, which was celebrated for several decades. Unfortunately, a particularly hard freeze in 1929, coupled with the onset of the Great Depression, led to its demise in 1930. The principal production of bulbs moved south to Skagit Valley, with its flat land, well-drained soil, and moist winters. After growers first began producing bulbs there in the 1920s, the industry took off in the 1940s, when Dutch immigrants arrived and prospered due to their Old World connections, and Skagit County became the leading producer of tulip bulbs in the country.

Rhubarb. Rhubarb, often referred to as the "pie plant," was a particularly profitable specialty crop for several farmers on Orcas and San Juan Islands. Around 1900 George Hershberger and

> ### *A Thousand Dollars per Acre*
> ### *This Is the Record of an Island Ranch*
>
> The possibilities of San Juan County soil, cultivated intelligently and with energy, are illustrated by the results obtained by John Lawson, of Deer Harbor, Orcas Island, the well-known rhubarb gardener. Mr. Lawson has been engaged in the cultivation of rhubarb for the past few years, supplying the Seattle and Bellingham markets, devoting an acre and a quarter to this crop.
>
> On this tract [in] the present year Mr. Lawson has raised twenty-seven tons of rhubarb, or 540,000 pounds, which he has sold at an average price of 21–22 cents per pound, bringing him on this acre-and-a-quarter tract the sum of $1,350.
>
> Many a man who thinks himself smarter than John Lawson pretends to be, complains that there is no longer room for a man of limited means in this world; but Mr. L. has demonstrated that a proper mixture of intelligence and energy will enable a man of limited means to derive a handsome income from a small tract of this desirable island soil.
>
> — *Friday Harbor Journal*, July 18, 1907

John Lawson, both of Deer Harbor, began growing rhubarb, which thrived on small, one- to two-acre plots of peat soil. Annual shipments of 10 to 20 tons fetched a price of three to five cents per pound—enough money ($600–$2,000) to provide a decent income for a farm family at that time. In 1909, the Island Packing Company in Friday Harbor reported shipping both fresh and canned rhubarb. At that time William W. Lee was also growing rhubarb in San Juan Valley. Although John Lawson continued to grow and sell rhubarb at his place in Deer Harbor into the 1920s, this crop probably stopped being raised commercially in the 1930s.

Ginseng. Ginseng, the root of plants in the genus *Panax*, has been used for millennia as a medicine, particularly among Asian cultures. During the early 1900s, the Bruns family raised both seed

[Dad] raised ginseng for 20 years, taking 7 [for it] to mature. The ginseng was all dried on racks in the kitchen and just the roots used. The more twisted and gnarled the better. It was used for medical purposes & was extremely bitter.

— Eleanor Bruns Still, Interview, 1990

and root ginseng on their Shaw Island farm for some 20 years. They had about three acres under cover and used maple leaves as mulch. The crop was sold to the International Ginseng Company in New York City as well as Chinese merchants in Vancouver, British Columbia. At one point, the Bruns family got as much as $11 per pound; however, during the year of their best harvest—a ton—they got only 75 cents per pound. In the 1990s, Jim Lawrence of Thirsty Goose Farm on San Juan Island raised ginseng as a cash crop.

Rabbits. Several domestic breeds of rabbit were brought to San Juan Island in 1903 by Messrs. Breedlove, H. Guard, and R. Guard. Some were released into the wild, but the severe winter

Harvesting Rhubarb at Hershberger Farm, Deer Harbor, Orcas Island

of 1916 killed most of them. In 1925, Howard Wilson and a Mr. Miller started a rabbit farm with a variety of domestic breeds on Cady Mountain on San Juan Island. Apparently, this venture did not work out, possibly due to the Depression, because the men released about 3,000 rabbits in San Juan Valley in 1934. By the 1960s, there were an estimated half million wild rabbits on the island, which posed a serious threat to farm crops. The government responded with a program of free fencing for commercial farmers. The rabbit population began to decline around 1979 and remains stable, with the largest concentration of warrens on the south end of San Juan Island. Rabbits, introduced from San Juan Island after their release into the wild, are also present on Lopez and Orcas islands.

Seed Crops. During the 1920s and 1930s, several farmers raised clover, vetch, and other soil-building legumes for seed. The 1925 census of agriculture reported 381 acres of red clover in San Juan County. A 1936 *Friday Harbor Journal* article estimated the annual production of vetch at 400 tons, with a price of $60 per ton; a record high of 1,227 acres was reached in 1940. During the 1940s, farmers experimented with several other seed crops, including table beets, cabbage, and spinach. Of these, only cabbage seems to have endured, at least until 1948. Small acreages of seed clover were also tried.

Scientific Farming

In America, the era of scientific farming had its beginnings in the Hatch Act of 1887, which allocated federal funds for land-grant colleges (founded under the Morrill Act of 1862) to establish agricultural experiment stations. These provided the basis for applying soil and plant sciences to farmland management and crop production. The agricultural experiment stations in turn became the foundation for the county extension system under the Smith-Lever Act of 1914. Farmers, who were often trying to improve their production through the application of modern science and technology, began to practice scientific farming through new breeds of animals and plants, soil management practices, and more efficient agricultural machines and farm structures. The scientific

> J. L. Davis, the well known pioneer farmer of Lopez island, went to Victoria Tuesday night with 274 boxes of apples. The shipment was made on the Hermosa, of Lopez, which cleared from here [Friday Harbor]. It is over forty-one years since Mr. Davis took his first shipment of produce from Lopez island to Victoria. That was five years before the settlement of the boundary controversy. There were no customs officers on the islands then and settlers going to Victoria with produce reported to Capt. Delacombe, in command of the garrison at English camp, and secured a permit from him.
> — *San Juan Islander*, December 12, 1908

approach was also applied to the design and layout of the farmstead as well as the farm household (home economics).

In the first half of the twentieth century farming in the San Juan Islands changed in response to these and other economic and social factors. Part of this change occurred in the area of transportation: getting crops to market and importing agricultural equipment and supplies. During the period of early settlement in the islands, homesteaders would paddle, row, or sail their goods, such as milk, butter, and eggs, to markets in nearby Victoria or Bellingham. They used Coast Salish canoes, EuroAmerican-style clinker-built rowboats and skiffs, and sailboats such as sloops and double-ended Columbia River gillnet fishing boats, both of which were easier to beach on the coves and small harbors of islands. Steamboats, which required docks, soon began routes between mainland ports and the islands, stopping at the many small hamlets and villages where island farmers could take their goods. This informal system of individually owned vessels, as well as several small companies of two or more ships, became known as the Mosquito Fleet, and served as an important transportation network for both freight and passengers. The first daily ferry to and from the San Juan Islands began in 1923. The small companies, as well as individually operated steamships, were consolidated by Joshua Green into the Puget Sound Navigation Company in 1913. In

> *The Steamer "Lapwing," of Victoria, cleared here for Victoria Sunday night with a cargo of sheep, lambs, peas, turkeys, chickens, apples and two horses—all valued at over $1,300, all purchased from San Juan farmers. Some of them are Populists but like their Republican neighbors they are indebted to these Gold Standard Republican times for the high prices and the good money that they received. There were 228 sheep and lambs in the shipment, making over 400 which the "Lapwing" has taken to Victoria from San Juan island this fall.*
>
> — *San Juan Islander,* November 10, 1898

1927 Green sold the PSNC to the Peabody family, who transferred their Black Ball flag to it, thus becoming known as the Black Ball Line. While individual shipping continued, the steamship system became the main means of transportation.

Several regional and local fairs promoted the islands' produce. In addition to the Alaskan-Yukon-Pacific Exposition, another major exhibition was the Northwestern Fair held annually in Bellingham from the late 1890s to the 1910s; Orcas Island fruit growers in particular exhibited their produce there. The first San Juan County Fair was held in a warehouse on the Friday Harbor waterfront in 1906. Local fairs were also held on several of the other islands; for instance, Eastsound boasted one in 1916. After fairs were held intermittently at various locations in the succeeding years, in 1923 local farmers issued stock for the purchase of land and construction of buildings near Friday Harbor for the San Juan County Fairgrounds. Several buildings were initially constructed: the Pioneer Cabin, a main exhibition building, and poultry and stock buildings. The first fair at the new site was held in October 1924; newspaper coverage mentions fruit, floral, and stock displays.

In 1913, the Washington State Legislature authorized the present extension system at Washington State College (later Washington State University)—a year before the US Congress passed the Smith-Lever Act, which established extension as a partnership of the US Department of Agriculture and the land-grant universities. With the success of farm demonstration railroad cars that toured east of the mountains, professors from the State College at Pullman planned the Farm Demonstration Tour of the islands in the fall of 1910 on a scow towed by a steamer, which would feature a model farm with an electrical light plant as well as spraying outfits, model poultry houses, a dairy herd, trees for demonstrations in pruning, and a "great variety of other farm and orchard paraphernalia." The tour, which occurred in August with reduced equipment and livestock (camping under canvas on the cramped scow, the crew only had room for one Jersey cow, for instance), stopped at Lopez, Olga, West Sound, and Friday Harbor, with interest on Lopez being principally on dairying and interest on Orcas on fruit culture. The first Washington State University County Agent to come to San Juan County, Russ Turner, travelled from Island County one week per month, starting in 1919. Two years later, William Ness became the first full-time county agent. Under Ness and his successors, farmers met to discuss improvement of their operations. Local and national publications helped in this effort. Both the *San Juan Islander* and the *Friday Harbor Journal* (1906–present) ran columns on farming matters; a regular feature in the 1910s was titled the Farm and Orchard Notes and Instructions from Agricultural Colleges and Experiment Stations of Oregon and Washington, Specially Suitable to Pacific Coast Conditions.

Several local organizations were formed for farmer cooperation and farm improvement. In 1899, George Meyers, George Gibbs, and George Adkins, with Ben E. Harrison as Secretary,

---

Agricultural Demonstration Boat

The equipment is considerably crowded upon a P.A. F. scow, covered with a canvas roof, the scow being towed from place to place by the tug Carlisle, of Bellingham. There is not room on board for many people at a time and the professors are obliged to dispense with nearly all of the comforts to which they are accustomed, taking their meals in somewhat primitive fashion and sleeping on the deck of the scow, in company with the Jersey cow used in the dairy demonstrations.

— San Juan Islander, August 12, 1910

Water Towers

Water towers were constructed to store water in a tank high above the ground, in order to provide water pressure through gravity. They were the sign of a prosperous farm, one that could afford modern amenities. Water towers, also called "tank houses," were usually built adjacent to a well, with an associated structure that housed the gasoline pump used to raise the water to the level of the tank.

Water towers were inevitably two or three stories high, where the top story usually held a round tank constructed of wooden staves with wood or metal hoops. Both the lower and upper stories could be clad in siding such as wood shingles, shiplap, or clapboard, although in a few instances the lower stories consisted of only the exposed supporting structure. When enclosed, the ground floor was sometimes used for farming operations, such as a milk room housing a separator; other ground floors were used for storage. When three stories tall, the middle, or second, floor was simply used as access to the tank above. The roof, covered in either wood shingles or metal, was often, although not always, pyramidal in shape. In most cases, the structure tapered from a larger footprint (12 to 17 feet square) on the ground to a smaller plan at the platform for the tank. In a couple of instances, a single-story structure was appended to the gable end of a two-story structure, which had vertical walls with no batter. In a singular case on Shaw, the 1938 tapered structure supports a top story that is octagonal, not rectangular, in plan.

incorporated the Orcas Island Fruit Growers Association to co-operate on marketing to Seattle, reducing prices on materials, and seeking reduction in freight charges. Later, the Orcas Island Commission Company was organized to help market products. In 1904, a farmer's cooperative built a warehouse in West Sound. In 1923, the San Juan County branch of the North Pacific Berry Growers Association boasted 27 members.

The Grange, established in 1867 as the Order of the Patrons of Husbandry, is America's oldest farm-based fraternal organization. A local ("subordinate") grange was first chartered with 43 mem-

Chas. Churchill is installing an engine and a complete electric plant on the farm of O. E. Clough. The residence, barn, chicken and other houses will be wired, and Mr. Clough will have every facility that a city power and lighting department could afford.

— *San Juan Islander,* November 15, 1912

bers on San Juan Island in 1908 as the Friday Harbor Grange #225. It was disbanded in the late 1920s and the San Juan Island Grange #966 was chartered in 1931. The Lopez Island Grange was organized in 1909, reorganized in 1914, and then a new grange (#1060) was organized in 1935. The Orcas Island Grange #964 was organized in 1931. In 1937, the San Juan Island Grange purchased a site on Spring Street for the San Juan Co-operative Market, providing grange members with access to refrigeration and freezer lockers, fuel, and scales for weighing crops and livestock. The cooperative was dissolved in 1946, although scales remained on site for grange use until 1963. The San Juan Island and Orcas Island Soil Conservation Districts were organized in 1939, after Washington State passed legislation that created the Conservation Commission. In 1947, the all-island San Juan Islands Conservation District was founded.

Prior to the establishment of the 1930s' New Deal Rural Electrifi-

Mrs. Waldrip's Water Tower

cation Administration, some farms had individual power generating systems. In 1937, twelve Orcas residents formed the Orcas Electric Company and under the renamed Orcas Power & Light Cooperative applied for a New Deal REA loan to construct a power station and transmission lines. Lines were energized on Orcas in 1939. In 1941, OPALCO bought L. T. Mulvaney's Friday Harbor Light and Power Company, and a year later bought a site for a power plant on Lopez. Power was initially generated by diesel engines; in 1951, a submarine cable was laid to the mainland to access the Bonneville Dam's generating facilities on the Columbia River. With the provision of electricity to the rural areas of the islands, agriculture in the San Juans entered modern times.

Mid- to Late Twentieth Century

After World War II, the agricultural landscape of the San Juan Islands began to change once again. Farmers, who had already struggled with marketing their crops off-island, found it increasingly difficult to compete with the mainland. By the time of the 1954 federal census of agriculture, agricultural labor constituted only 28.2 percent of the county work force. Large-scale, sole-income farming began to disappear.

Changes in agricultural technology contributed to this. Tractors were introduced to the islands as early as 1924, but most farms continued to use draft horses prior to World War II. After reaching a high of 1,102 in 1920, the number of horses and mules in San Juan County declined in the next several decades but was still at 499 in 1940. After the end of the war (1945), however, the number dropped precipitously to 288, and by 1954 only 85 horses and mules remained.

The local dairy industry was affected by consolidation of operations on the mainland. A group of five Puget Sound dairy co-ops joined together in 1918 to form the United Dairymen's Association; in 1930 this group bought Seattle-based Dickey's Consolidated Dairy Products to form the Northwest Dairymen's Association, a co-op with nearly 450 farms throughout Washington State that still sells products through their subsidiary Darigold. In 1949, the state legislature passed the Fluid Milk Law (also known as the Grade A Milk Ordinance), with more stringent regulations than before. Larger operations were able to meet these requirements more easily than smaller ones such as those on the islands. Dairy cattle, some 3,000 strong on the islands in 1954, declined to 372 just five years later. The creamery in Friday Harbor closed in 1962.

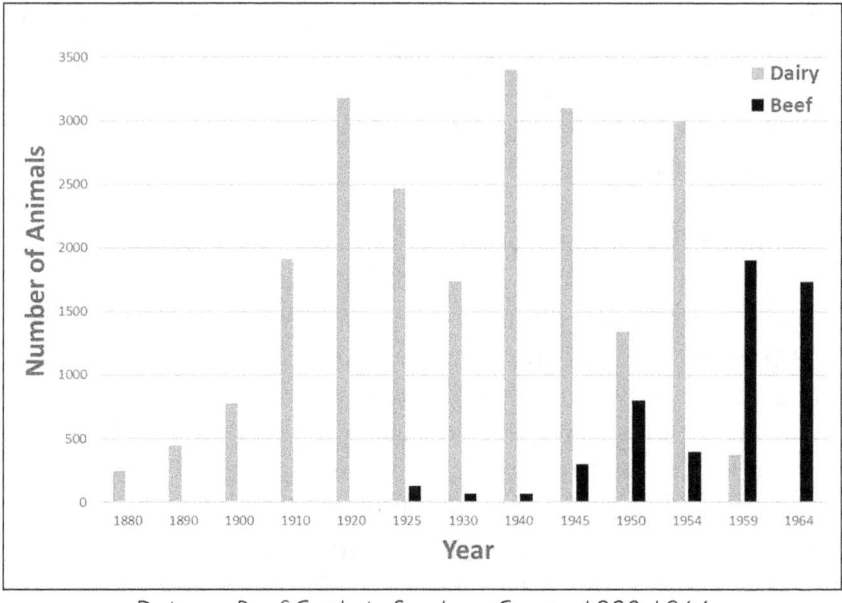

Dairy vs. Beef Cattle in San Juan County, 1880-1964

During the same period, beef cattle—less labor-intensive and not as heavily regulated—rose from 400 on the islands in 1950 to almost 2,000 four years later and have fluctuated from 1,200 to 2,000 during the years since then. At first, in the mid-1940s, stockmen preferred a beef-dairy cross, but after World War II Orcas farmers introduced purebred Herefords, which gained in numbers. In 1960, Fred Zylstra began raising purebred-registered polled Herefords, principally as breeding stock, at his Wooden Shoe Farm in San Juan Valley. He held auctions at the farm, where bidders would come to buy his bulls. Most beef cattlemen eventually settled on Herefords and Black Angus, with a few specializing in Charolais (Charles Arnt's Driftwood Ranch on Orcas Island) and Scotch Highland (Dana McBarron's Cape St. Mary Ranch on Lopez Island). However, more stringent US Department of Agriculture regulations regarding slaughtering and butchering led to the closure of local slaughterhouses and meat markets; island butchering survived for local consumption while market beef was shipped off-island to USDA-approved slaughterhouses.

Peas as a major crop were revived in 1956, when George P. Jeffers and his associates who ran the fish cannery in Friday Harbor bought San Juan Valley Farms and grew about 450 acres of peas yearly. According to Tony Surina, who worked in the fields, the land was plowed and disked at several angles, fertilized, and disked again. The seeds were planted with a drill, which could cover 10 acres per day; the soil was then smoothed with a float or drag. When the weeds emerged, the plants were sprayed with "Pre-merge," or sometimes crop dusters flew over the valley. Harvest occurred around the Fourth of July, when the pea vines were cut and winnowed with an Oliver tractor while the pea viners—this time mobile, in contrast to the older, stationary ones—separated the peas, which fell into boxes that were then hauled into town for canning. In 1958 and 1959, the fields were planted in oats and barley to build the soil. Jeffers built the Friday Harbor Canning Company, a freezer operation, in 1960. In 1961 and 1962, 450 acres of peas were planted, with the peas frozen in 50-pound totes at the plant, which ran 18 hours a day. In the last year of production—1966—485 acres were planted, yielding 700 tons. The economics of large-scale commercial farming, even with a crop as profitable as frozen peas, didn't pencil out.

Several studies in the mid-1950s help to form an overall picture of farming in the San Juan Islands during the mid-Twentieth Century. The first of these studies occurred from 1952 to 1953 when several students studied the agriculture of San Juan Valley for a field school as part of a geography course at the University of Washington. They surveyed farmers and conducted research on historical trends in crops and farming methods. The students noted the preponderance of livestock farming, particularly as a source of market income, with the main emphasis on sheep raising, but which also included dairying and beef cattle: at the time, 31 farms were selling dairy products, 22 were selling sheep or wool products, and 19 were selling beef. Grains—wheat, barley, and especially oats—were also important crops, but largely for internal consumption rather than the market: only seven farms sold hay or grain. A paper by Henry H. Davis, "Part Time Farming on San Juan Island," written in August 1952, noted the changes in the economy of farming in the islands following World War II, emphasizing the growing trend towards seeking jobs off of the farm. In terms of irrigation, the Farm Water Supply Census, compiled by George H. Smart in the summer of 1953, noted 26 existing farm ponds, with four planned and seven irrigation systems: four sprinklers, two pipelines, and one ditch. A summary map titled Land Uses of San Juan Island from July 1953 indicates that the main area of crop lands on the island was centered in San Juan Valley, with outlying areas near American Camp, Beaverton Valley, and West Valley on the north end. A more detailed map of San Juan Valley by Duilio Peruzzi, also compiled and drawn in the summer of 1953, indicates large areas of forage or pasture as well as some specific fields of grain.

A second important study was the 1956 San Juan County Agriculture Study, part of the *County Agricultural Data Series* published by the Washington State Department of Agriculture. Based on the 1954 federal census of agriculture, the study includes additional data on the history of farming in San Juan County as well as a comparison of past census information. In noting that agriculture played a smaller role in the county's economy—28.2 percent of workers, still the largest group, followed by workers in manufacturing (18.4 percent), miscellaneous services (17.1 percent), retail and wholesale stores (10.5 percent), and construction (9.1 percent)—it noted that the value of crops mainly resided in livestock, particularly beef cattle. Of the total value of county farm products in 1954 ($641,240), 27.6 percent was in cattle and calves, compared to only 11.4 percent for dairy products, milk and cream, and 6.6 percent for grain and hay crops. While the number of beef cattle in the county was less than the number of dairy cattle (1,900 vs. 2,000), the value of the beef stock was higher, making up 35.4 percent, compared to 33.2 percent for milk cattle of the total value of livestock in 1954. Of particular interest in the study is the section on irrigation facilities, which noted that there were only five farms that irrigated their land in 1954 (down from seven in 1950), with a total of 80 irrigated acres—44 in crops and 36 in pasture.

In the 1960s the Soil Conservation Service offered design assistance to landowners who wished to construct ponds for both irrigation and wildlife preservation. The *Friday Harbor Journal* featured stories on the construction of several ponds, including a 1966 article entitled "Man-Made Ponds Popular in San Juans," which stated that there were "approximately 200 man-made ponds" in the islands and concludes with "landowners in San Juan County are encouraged by the local Soil and Water Conservation District to consider more water retention on their property."

Construction Underway on Wooden Shoe Farm Dam

Tentatively scheduled for completion by the first of September, the dam will flood 48½ acres of land and provide supplemental irrigation for 80 acres of cropland. The structure will be approximately 300 feet long, have an embankment fill of 8,254 cubic yards, a maximum height of 17 feet and store 288 acre feet of water.... When completed this earth structure will cause formation of the largest man made lake in the State of Washington built by a private individual.

— Mike O'Keefe, Soil Conservation Service
Friday Harbor Journal, August 15, 1963

The largest pond was Zylstra Lake in San Juan Valley. Planning for and construction of the dam for the lake began in 1963. Construction was made possible with technical engineering assistance through the Agricultural Conservation Program for wildlife conservation practices with soil and water benefits of the SCS, locally administered through the San Juan Soil and Water Conservation District No. 43. Among the ponds and lakes built in the islands with assistance from the SCS were those for Bud Wold (in 1963), Eric Erickson (in 1964), and Gordon and Lois Jorgenson (in 1966), all on San Juan Island, and Fowler's Pond (in 1963) on Orcas. A 1975 report on the geology and water resources of the San

Juan Islands emphasized that Zylstra Lake was the most important irrigation water reservoir in San Juan County, on the average supplying about 300 acre-feet of water and irrigating about 400 acres of farmland in San Juan Valley.

In 1962 the SCS of the US Department of Agriculture, in cooperation with the Washington Agricultural Experiment Station, published the *Soil Survey [of] San Juan County, Washington*. The fieldwork had been done in 1956, and aerial photos were taken in 1960. The report not only identified soil groups, characteristics, and capabilities of farmland throughout the islands, but also offered general information about the county in terms of physical characteristics, history and development, and agriculture specifically, based largely on the 1954 federal census of agriculture. A specific mention is made in the report about water supplies: "Supplemental irrigation is used by a few farmers, primarily during July and August. The water for irrigation is taken from lakes and small reservoirs on the farm. Only a small part of the acreage used for agriculture is irrigated, but the size of irrigated acreage increases each year."

Into the Present

With the growing attraction of seasonal residents and retirees to the islands beginning in the 1960s, the year-round population of the islands, after remaining relatively flat around 3,000 from 1910 to 1970, jumped to 7,838 in 1980 and climbed rapidly to 15,769 by 2010. The cost of farmland rose dramatically, in pace with residential and commercial land costs. During the 1970s, San Juan County went through the process of adopting a Comprehensive Plan; as part of this plan, some areas of the islands were zoned as Agricultural Resource Land, a distinction based upon the delineation of prime soils as defined in the 1962 soil survey. According to the December 2017 *Economic Analysis of Resources Lands*, there are currently 3,900 acres of ARL—about 89 percent of the estimated 15,700 acres—being actively farmed.

During this same period, San Juan County implemented the 1970 Washington State Open Space Taxation Act, which allowed qualified property owners to have their agricultural land valued at current use rather than at highest and best use. As of 2017, there

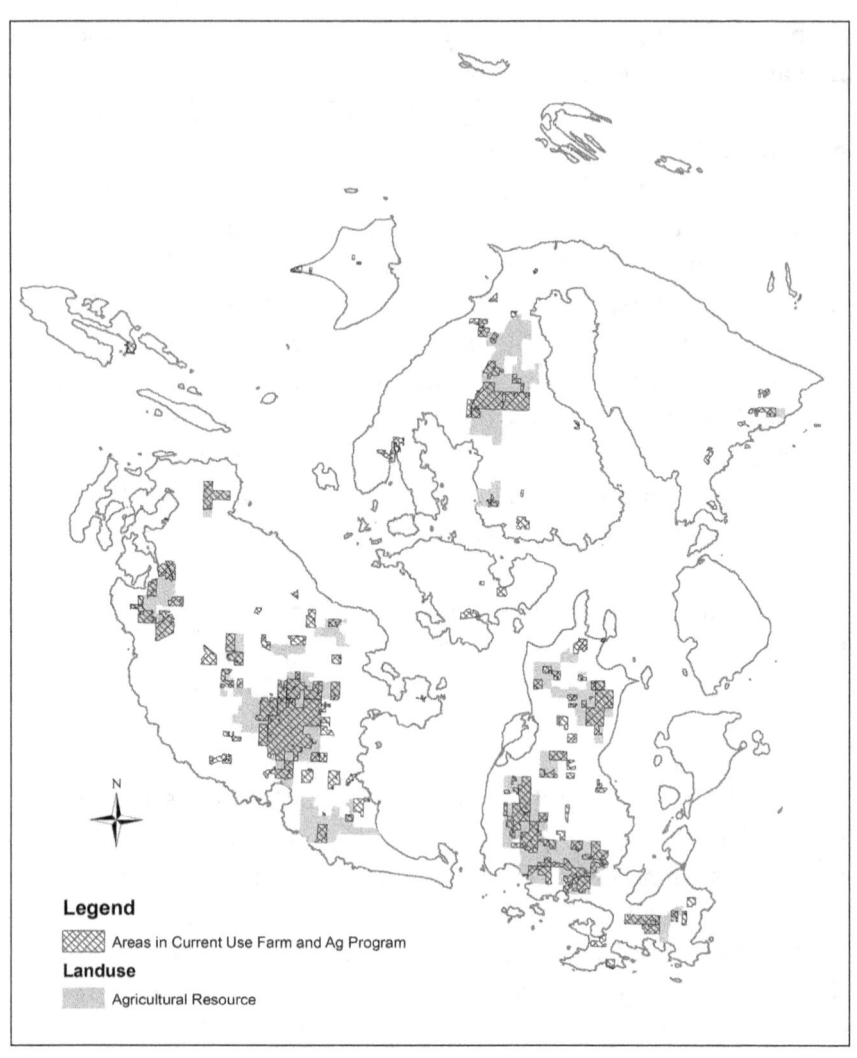

San Juan County Agricultural Resource Lands and Current Use Farm and Ag.

were 10,086 acres listed as "Open Space–Agriculture" in the Current Use Farming and Agriculture Program administered by the San Juan County Assessor. In 1990, the Washington State Legislature passed the Growth Management Act, which, among other things, led to the designation of critical areas and natural resource lands; this strengthened the protection of agricultural resource land. Two organizations currently help conserve farmland in the islands: the San Juan Preservation Trust, a nonprofit land trust established in 1979, and the San Juan County Land Bank, a public organization established by vote in 1990 and funded through a real estate excise tax. By 2018, the Preservation Trust had protected 2,925 acres of conserved farmland (2,100 in easements and 825 acres in preserves) and the Land Bank had conserved 1,700 acres of farmland (1,121 in easements and 579 in preserves).

Within the last thirty years, agriculture in the San Juan Islands has been on the rebound, albeit in different forms from the past. The most recent data is from the 2017 federal census of agriculture. After reaching a nadir of 113 in 1974, the number of farms in San Juan County grew from 2002 to 2007 (225 to 291, or 29 percent), fell slightly in 2012 to 274, and then rose to 316 in 2017. Total acreage also grew between the 2002 and 2007 censuses: 17,146 to 21,472, or 25 percent, although it then fell significantly to 15,669 in 2012; in 2017, it grew to 18,402. Most of this growth was in small farms: farms 1 to 9 acres in size grew steadily from 32 farms in 2002 to 42 farms in 2007, 49 in 2012, and then jumped to 72 in 2017; those farms 10 to 49 acres in size grew from 119 farms in 2002 to 155 farms in 2007, and then dipped slightly to 151 in

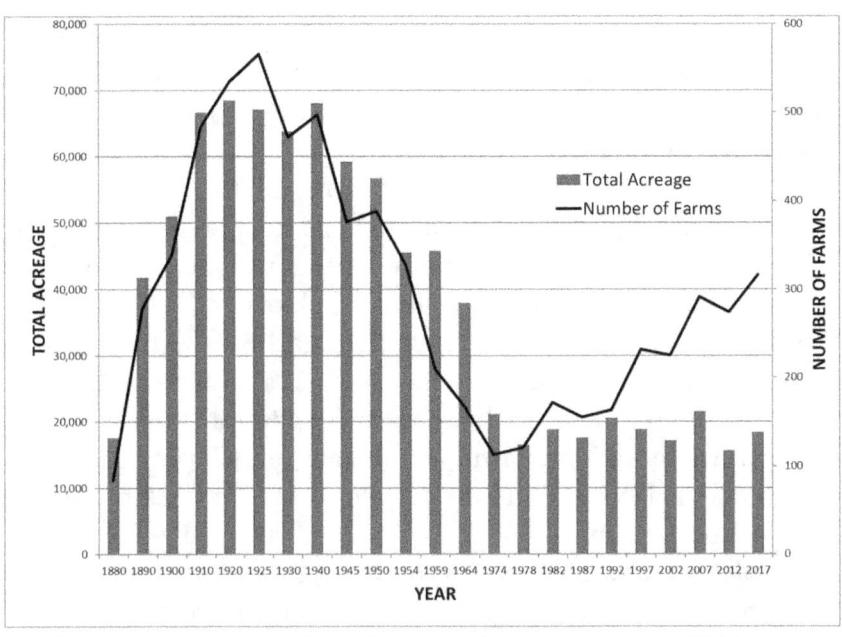

Total Acreage and Number of Farms in San Juan County, 1880-2017

2012, and remained at that number in 2017. Two interesting statistics stand out in the recent census data: the average age of San Juan County farmers was 60 and there was a higher percentage of women producers (50 percent) in the islands than in the state (42 percent) and the country (36 percent).

Currently in terms of land area, Lopez Island, with its open rolling hills and loamy soils, has the largest extent of agricultural activity: 4,967 acres, or 26 percent of its land mass (18,900 acres). San Juan Island, with its mixture of open fields and forested mountain areas, is second, with 6,140 acres, or 17 percent of its 35,500 acres. (San Juan Valley—the False Bay Creek watershed—accounts for 59 percent of this number.) Orcas Island—the most mountainous—has only 2,255 acres, or 6 percent, of its landmass (36,900 acres) in agricultural use. Waldron, although much smaller in size—2,900 acres—has significant farmed land: 257 acres or 9 percent, as compared to Stuart (7 percent) and Shaw (4 percent).

Since 1990, agricultural employment in San Juan County has grown by 39.5 percent, with an annual average of 1.3 percent higher than in Washington State (2.4 percent) and 0.7 percent less than that in the United States. There were 215 agricultural jobs in the islands in 2015, with an average of 262 during the period of 2001–2015. This represents 2.3 percent of overall current employment in the county, but its economic influence extends to areas such as real estate and tourism.

Even though flocks declined after the 1950s, sheep still constitute a sizeable portion of the farming landscape. After reaching a high of 3,013 in 2007—the largest number per county in western Washington—the number had fallen to half of that (1,583) in 2017. The other major livestock—beef cattle—reached a high of 3,420 in 1997 and stood at half that (1,706) in 2017. As marketable crops, there is a trend away from livestock raising toward vegetable production: while the total farm gate in San Juan County rose from $2,823,000 in 1997 to $4,119,000 in 2017, the livestock portion of that declined—from $2,366,000 to $1,705,000—while return from vegetable crops rose from $457,000 to $2,414,000.

There has also been a growing trend in two areas: direct sales and organic farms. The federal census of agriculture first began recording the "value of agricultural products sold directly to individuals for human consumption" in 1997; direct sales in San Juan County climbed from $174,000 in that year to $418,000 in 2002, $739,000 in 2007, and reached a peak of $1,170,000 in 2012, but then fell to $743,000 in 2017. Within the last three decades, farmers markets have emerged at Friday Harbor (San Juan Island), Eastsound (Orcas Island), and Lopez Village (Lopez Island), and there are numerous farm stands on each of the islands.

According to the census, which first began recording data on organic practices in 2002, the number of farms that have been certified as organic in San Juan County rose from 11 in that year to 32 in 2007, and then dropped to 16 in 2012 and 12 in 2017. Sales of organic produce jumped from $115,000 in 2002 to $288,000 in 2007, and then dropped to $220,000 in 2012 and $198,000 in 2017. One reason for these recent declines may be that the added expense associated with organic certification is perceived as unnecessary: most of the produce is sold locally to consumers who are aware of farms' practices and don't need to be reassured by certification that they are using organic methods.

Orcas Island Farmer Vern Coffelt, 2007

An interesting aspect of the economic geography of the islands is illustrated by the case of the farmers on Waldron, one of the northernmost of the San Juan Islands, which is not connected to the power grid supplied by OPALCO. Waldron farmers produce their crops and either ship them to the farmers markets on Orcas and San Juan or freight them by UPS and other package services to markets throughout the nation. To prepare for the farmers markets, they harvest their produce the night before, keep it cool overnight in cedar groves, and then load their boats early in the morning in order to make the mid-morning opening times.

Early Farmers Market at the American Legion in Friday Harbor

Island farmers today specialize in crops such as alpacas and llamas, lavender, herbs, and value-added products such as jams, jellies, and sauces. Thirsty Goose Farm on San Juan Island is a good example of an operation that has adapted to the vicissitudes of the market. Started by Jim and Lisa Lawrence in 1974, Thirsty Goose crops have included beef cattle, sheep (for both food and fiber), ginseng, asparagus, and tomatoes. Recently, the Lawrences have transferred their operations to the next generation, which has continued with greenhouse production of tomatoes and mixed salad greens.

Coffelt Farm, on Orcas Island, has a long history of cultivation. Located at the upper end of Crow Valley, the land was farmed by Vern and Sidney Coffelt. After 60 years of farming, the Coffelts sold their property to the San Juan County Land Bank to help realize a mission of preserving working farmland. In 2011 operation of the farm was transferred to the nonprofit Coffelt Farm Stewards, whose mission is to "demonstrate sustainable, island-scale agricultural practices, promote environmental stewardship and provide opportunities for education and research while honoring Orcas Island's rural heritage." The Stewards and their lessees produced grass-fed beef, pastured pork, and grass-fed lamb; broiler chickens and eggs; fruit; and wool products. Recently, Lum Farm, LLC has leased the farm and are milking goats as well as offering eggs and broiler chickens, berries, rhubarb, garlic, orchard fruit, beef, lamb, goat meat, walnuts, sheep and goat skins, and hay at the farm stand.

Sunnyfield Farm is a new operation on an old farmstead: the Harry Towell Farm on Lopez Island. Andre and Elizabeth Entermann began a goat dairy operation on Elizabeth's parents' place, and in November of 2014 the Washington State Department of Agriculture granted them a license to sell raw goat milk, aged cheese, and pasteurized cheese. Sunnyfield Farm sells its products to local stores and restaurants, as well as from a farm stand and at farmers markets in Friday Harbor and Lopez Village.

There are currently four wineries in the islands: Lopez Island Vineyards (est. 1987); San Juan Vineyards (est. 1996); Orcas Island Winery (est. 2011); and Madrone Cellars & Ciders (est. 2017), all of which grow white wine grapes such as Madeleine Angevine and Siegerrebe and import red grapes from the Yakima and Columbia valleys for their red wines. In the mid-1990s, Westcott Bay Cider planted apple trees in an orchard on White Point on San Juan Island and produced its first cider in 1999. Eleven years later, new partners joined to form San Juan Distillery, which makes gin, brandy, and other products from the apples and naturally sourced botanicals. In 2014, Orcas Island Distillery began making pear and apple brandies from fruit picked from the island's historic orchards. Similarly, Girl Meets Dirt on Orcas Island sources fruits from historic and modern orchards to produce its line of bitters, shrubs, jams, and preserves. Madrone Cellars & Ciders uses hard-cider apples from Bellevue Farms on the north end of San Juan Island as well as imports apples to make special-blend and single-varietal hard ciders.

The first Community Supported Agriculture program—where customers subscribe at the beginning of the growing season to weekly deliveries of vegetable and livestock crops—was established on San Juan Island in 1992. In 2001, the first-in-the-nation USDA-inspected Mobile Slaughter Unit was created by Washington State University San Juan County Extension and the Lopez Community Land Trust; the Island Grown Farmers Cooperative was then established to operate it. This gave producers the ability to slaughter their livestock on the farms and, after the carcasses were taken to a facility for butchering, to sell them by the cut. There are currently two certified dairies in the islands: Our Lady of the Rock on Shaw and Sunnyfield Farm on Lopez. In 2005, San Juan County established the Agricultural Resources Committee; the same year, the Northwest Agriculture Business Center, serving Island, King, San Juan, Skagit, Snohomish, and Whatcom Counties, was formed as an economic incubator.

Although it is not considered an "agricultural product" by Washington State, marijuana production arrived in San Juan County after the passage of State Initiative 502, On Marijuana Reform, in the 2012 general election. The State both licenses and regulates the production and processing of marijuana, requiring among other things production either in enclosed and secured greenhouses or outdoors within a fully enclosed physical barrier, with a sight-obscure wall or fence at least eight feet high. As of 2019, there are seven licensed producers in the islands; of these, only two are currently reporting monthly sales.

New technology has helped farmers farm their land more efficiently as well as allowing for farming on smaller acreages. High-tensile electric fencing began to replace the standard barbed wire and metal mesh fences, allowing for reduced cost for materials and labor, less maintenance, and extended coverage. Drip irrigation, with pumps and lines running from ponds or wells, has helped reduce water use. Rotational grazing methods have led to more intensive and regenerative use of pastureland. Technology such as hoops and greenhouses has allowed farmers to extend their growing season. In 2005, the SCS conducted the updated, more accurate and detailed Soil Survey of San Juan County, which has assisted farmers in identifying and responding to specific soil conditions in their fields. Recent concerns about carbon sequestration—the retention of carbon in the soil to help mitigate or defer carbon dioxide buildup leading to global warming—have led to local experimentation with no-till (also called zero tillage or direct drilling) methods of farming, where crops are grown or pasture is maintained from year-to-year without disturbing the soil through tillage.

Today agriculture is a significant part of the economy of the San Juan Islands. Although it does not have the statistical heft of real estate, tourism, and construction—the three major drivers of the islands' economy—it contributes to the beauty of the islands' landscape, the lived experience of visitors and locals alike, and the sustenance of our rural economy. Island farming continues to adapt, as it has in the past, to climate and the marketplace.

What about that Mink Farm...

Beginning around 1940, John Jackson farmed mink on Egg Lake Road on San Juan Island. He fed the mink fish guts from the cannery in Friday Harbor, which added a glossy sheen to their coats. Jackson sold the mink farm to Evert and Bertha Hall in 1946, the same year he bought property for a cannery south of Friday Harbor. Local sources say that the operation continued, with feed from Jackson's new cannery, until at least 1957, possibly as late as 1970.

INTER-ISLAND FERRY ROUTE

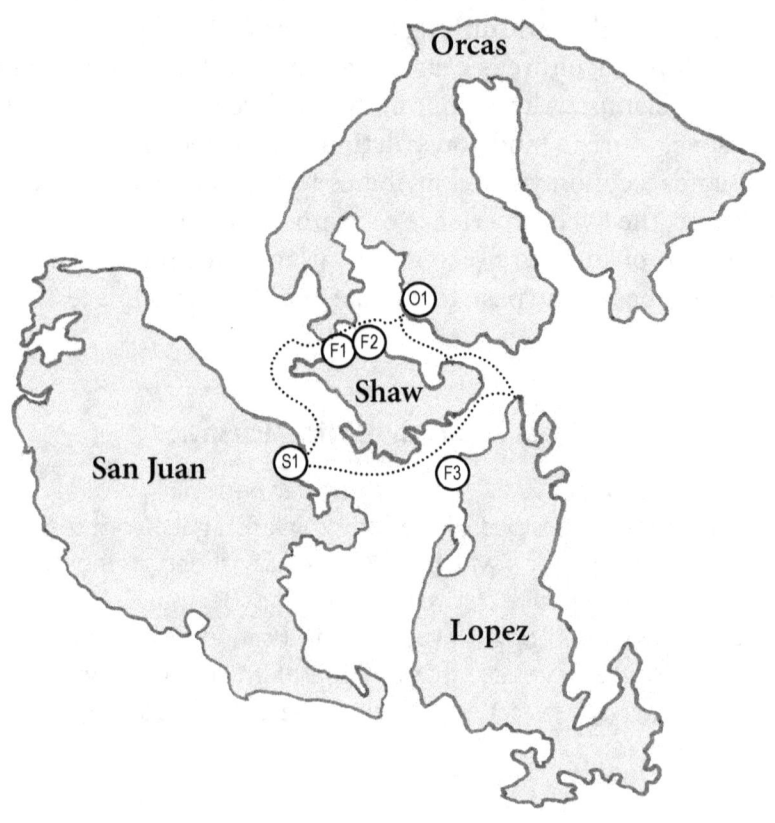

TOURING THE AGRICULTURAL LANDSCAPE OF THE SAN JUAN ISLANDS

> *The beauty that we see in the vernacular landscape is the image of our own common humanity: hard work, stubborn hope, and mutual forbearance striving to be love.*
>
> — John Brinkerhoff Jackson

You can see a good portion of the agricultural history of the San Juan Islands while traveling around the islands and observing the landscape. From the public roads you can view farmsteads including fields, fences, and drainage ditches. Visible farm structures include houses, barns, granaries, milk houses, silos, and water towers. By examining the location, arrangement, shape, and design of these buildings, you can often infer the primary crops and how they contributed to the workings of the farm. The most dominant feature—both in terms of size and significance—of most farms is the barn; its importance is often recognized by the Washington Heritage Barn Register.

Each of the major islands in the San Juans is unique in its agricultural environment. Geologic, geographic, and climatic factors generated different farming spaces and conditions, which in turn influenced the types of crops raised and how farmsteads were laid out. Orcas Island, for instance, is very hilly, with concentrated pockets of farmland in places like Crow Valley and Eastsound; Lopez Island, in contrast, is relatively flat and has abundant farmland distributed island-wide. San Juan Island is a mixture of the two, with a large concentration of agriculturally rich land in San Juan Valley and pockets of arable soil in Beaverton Valley and West Valley.

An important factor in agriculture—both subsistence, or trade, farming and market farming—is transportation. Trails and roads offered routes to transport goods and supplies to and from other farms, villages, and ports. The 1874 township and range survey of the islands indicates that early roads took advantage of topography—following contours, crossing hills and mountains through passes, and fording streams in shallow spots—to connect farms in the most direct manner. Later, with the imposition of the Public Land Survey System, farms were based on homesteads defined by sections and their orthogonal divisions. Roads, which straddled the boundaries of each homestead along the east-west and north-south section lines, form doglegs that don't necessarily conform to the lay of the land. However, this is only possible on relatively low-profiled topography, meaning that it is rare on Orcas, more numerous on San Juan, and widespread on Lopez.

From these roads you will first see fields of various crops—predominantly hay and pasture—as well as the fences that enclose them. The roadbed itself is generally elevated to avoid flooding during the rainy season; bar ditches line the roadways to drain the water off. Fences—today predominantly post and wire mesh (although there are some electrically charged)—line the fields abutting the roads to prevent farmstead animals from getting loose and prevent neighbors' animals from intruding. Fences also subdivide fields, and ditches drain these subdivisions.

Farmers usually built their houses and agricultural structures close to roads so that they could access them easily. With the advent of the homestead system and location of roads along section lines, farmsteads were sited farther from the road. Topography also dictated the location of farm buildings. Farmers didn't build structures on prime arable soil; instead, they picked rocky, well-drained outcroppings for ease of access. Furthermore, these were often located at tree lines and on ridges or rises overlooking (surveilling) the farmers' fields. At first, this was a EuroAmerican pioneer response to the prevalence of raids by North Coast Indigenous tribes such as the Haida and Tsimshian. Later, it facilitated an efficient means of observing livestock and crops.

In order to explain the farm structures visible today, this tour guide may repeat portions of the history part of this book, because while it is meant to supplement the history, it can be used independently. The sites are ordered along routes starting from the ferry dock of each island. Street addresses are only given for sites that may be publicly visited. At some sites no historic structure remains, but the site location itself is significant enough to warrant mention. In the identification of each site, the historic names, where known, are given first, with current names in parentheses.

This tour guide presents the agricultural landscape of the islands that can be observed from the public right of way. Please respect the fact that most of the land you will be looking at is private property; unless there is a farm stand or other venue clearly advertised as open to the public, do not trespass. Such publicly accessible venues are indicated on the maps and text in this section, but hours of operation are subject to seasonality and other factors, so please check before you visit. Unfortunately, there is no guarantee that a structure such as a barn will still be standing when you visit; the agricultural landscape is, by definition, one of change.

Inter-Island Ferry Route

Farms seen from the ferry. Several farms can be seen from the inter-island ferries. In addition to general farm landscapes—fields, fences, and some farm structures—the site of once-extensive GEM Farm on the north end of Lopez is clearly visible, but no structures remain from the time of its operation. On the route between Friday Harbor and Orcas, as you approach Broken Point on the north side of Shaw Island, you'll catch a glimpse of the barn and the silo at the Biendl Farm and the barn at the Shaw Farm.

F1: John Biendl Farm. John Biendl arrived on Shaw in 1902 and married Ruth Ellen Shaw, who was born on the island. Their farm, which has a large dairy barn, is one of the few in the islands with an intact **silo** (1919). The gable-roofed **barn** (listed on the Washington Heritage Barn Register), which measures 50' wide by 54' long, and is 31' high, is built into a slope, so that part of the floor was earth and part raised-wood platform. Like many farmers in the islands, John Biendl did a variety of jobs to make a living: for instance, he sold cordwood for burning in the lime kilns at Roche Harbor. Scows were loaded from the shore below the barn. In the past, the barn's owners have put up a wreath on the barn that fills the early dark evenings with warm light—a beacon for winter-weary ferry travelers!

F2: Shaw Farm. Richard J. Shaw—not the person for whom Shaw Island was named—arrived on the island in the 1890s and married Janette Ellen Jennie Gordon. In addition to a large barn, he built several farm-related structures, including a chicken coop, milk house, smokehouse, outhouse, and root cellar. The **barn** measures 20' wide by 32' long, with a 16' shed, and is 24' 6" high. The main gable area has a hay hood and door, with wood rail and steel trolley for the mow. The shed addition shows evidence of stanchions and a manure gutter, indicating that the structure was used as a dairy barn. Downslope is a cluster of farm buildings, some of which can be seen from the ferry. The 13'-by-24' **chicken coop**—enough for a couple dozen birds—is constructed with vertical 1x12s nailed to sills on the ground. The white **milk house** is a typical 6'-by-8' gable-roofed-structure, sided in clapboard. A "two-holer" **outhouse** is located nearby. A small—3' by 4' by 5' 6"—**smokehouse** (most smokehouses are larger—5' by 6'—and taller—8' to 10') has a "ladder" inside that was probably used for hanging meat or fish. Perhaps the most interesting structure on this site is the **root cellar**, which was built into the hillside with square-notched, sawn logs; dirt-and-rubble fill was used to insulate the structure.

F3: GEM Farm. There is little left of the eight-hundred-acre GEM Farm, established by Ben Lichtenberg on Lopez Island in 1898, which produced several crops, including fruit such as apples, pears, plums, and cherries; poultry such as Plymouth Rock cockerels and White Holland turkeys; and dairy from Jersey cattle. Compare this view of Lopez Island with the photograph of GEM Farm at its height (page 57).

O1: Orcas Landing. See page 151.

S1: Friday Harbor. See page 161.

LOPEZ

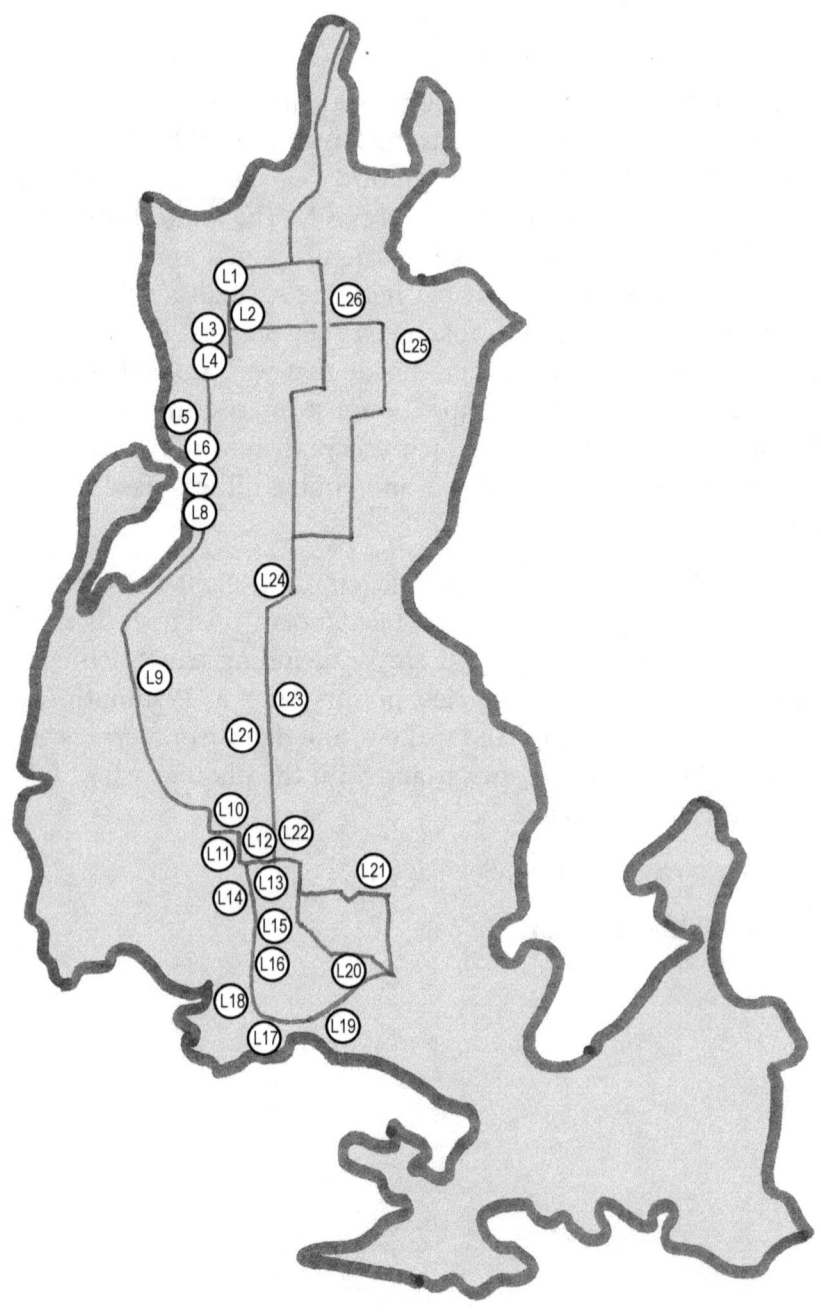

Lopez Island

> *Lopez Island has an area of forty square miles . . . and of all the islands of San Juan County presents the largest proportion of tillable soil. The valleys of this island embrace acre after acre of the most fertile land under the sun and charm and delight the eye with their beautiful fields and well-kept, attractive orchards. The island is comparatively level and there are no water lands of any consequence. . . . From north to south it is one grand valley, and were it all cleared up the view would be one of a large garden.*
>
> — *The San Juan Islands, an Illustrated Supplement to the San Juan Islander,* 1901

Agricultural land on Lopez is somewhat evenly distributed, although there are distinctly named farming areas, such as Lopez Village, Center Valley, and Richardson. The ferry dock is at the wooded north end of the island, where there is no farmland. There is evidence of Coast Salish farming—principally camas beds—on portions of south Lopez, and Lopez Village is on the site of an old village, but all remaining sites date from the post-contact period when EuroAmericans established homesteads that later developed into large commercial farms.

L1: Roy Prestholt Turkey Barn. As you proceed south towards Lopez Village and turn the dogleg corner near Military Road, you encounter a large, monitor-roofed building with a bicycle in the gable. In the 1950s Roy Prestholt built this barn for raising turkeys on land owned by T. J. Blake; allegedly it went up in a mere three weekends. The barn was designed to house poultry in pens on the ground floor, while allowing access for feeding and ample ventilation though a central draft system. However, it has also been said that the barn was never used for its original purpose. Oriented east-west, the structure is 60' wide by 100' long by 27' tall, and has bands of windows along its sides to allow for plenty of light for the birds. The monitor walls also have ventilation slats, allowing for ample ventilation, and there are two bands of windows on the gable ends for more light. The plan of the building consists of two 20-foot-long banks of pens flanking a 20-foot-wide center drive for delivery of feed and removal of waste. The barn is constructed with a concrete floor and stem walls, with wood framing above. A prominent feature of this barn is the large gable dormer on the north side, which was used as a granary for poultry feed; the dormer doors allowed for loading the feed directly into the second-story space above the pens on the ground floor.

L2: Thomas G. Blake Farm (Lopez Island Farm) (193 Cross Road). The farmland from the dogleg at Military Road on the north to the dogleg on the south was formerly owned by the Blake family. Patriarch James Blake, who was born in Ireland in 1827 and emigrated with his family to Canada when he was three years old, came to Lopez Island in 1883, and bought a quarter section of land comprising the core of this area. His son Thomas George obtained this property near the intersection of Cross and Fisherman Bay Roads in 1897 and made it his farmstead, where he raised a variety of crops.

The farm buildings on the site—one of the more complete agricultural complexes in the islands—include a house, barn, chicken coop, machine shed, well house, and the foundation of a silo. The **house**—closest to the road and the first structure you encounter—has been added onto several times, but the oldest portion consists of a side-facing, two-story gable-roof structure with a recessed porch along the south side. Beyond this lies the cluster of farm buildings. The timber-frame **barn** is built with a center drive for delivery to the haymows on either side. It is 31' 3" high—allowing for plenty of loose hay—and measures 34' 6" wide (and 50' 2" long)—allowing for the center drive and the flanking mows. The **chicken coop**, measuring 12' wide by 21' long (probably at one time housing several dozen birds), has been converted to a certified kitchen and farm store, where you can purchase Lopez Island Farm products, including grass-fed lamb and pasture-raised pork. The **machine shed**, roughly built with wood sills on fieldstones and vertical wood boards for siding, measures 16' by 18'—enough room for several farm machines. The 8'-square **well house** may have been a milk house, with concrete foundations and floor indicating a cool, clean, washable space for separating and storing the cream. And although only the foundation of a 12'-diameter **silo** remains—typical among farmsteads these days—it does indicate that the farm was a dairy operation, because the silage stored in the silo would have been fed to the dairy cattle.

L3: Lopez Island Vineyards (724 Fisherman Bay Road). Lopez Island Vineyards (est. 1987) is the oldest of the four vineyards in the islands. Vintner Brent Charnley, together with his wife Maggie Nilan, grows organic white wine grapes such as Madeleine Angevine and Siegerrebe and produces several wines from his own varietals and from grapes from eastern Washington. The pressing and tasting rooms are surrounded by the vineyards of various grapes. You can visit the tasting room during seasonal hours.

L4: James Blake Farm. Heading south, you encounter a vista of the former James Blake Farm before you round the corner and afterward on your right. The principal structure here—a timber-frame barn—was probably constructed in the 1890s by James Blake, who bought the property in 1883; it was later owned by his grandson, T. J. Blake. It measures 40' wide by 50' long and is 31' to the ridge—allowing enough room for a center drive to access the flanking haymows, where the loose hay was piled high. There is a shed addition on the southwest side. The newer **milk house**, 10' wide by 12' long, with a 4'-high concrete stem wall, indicates that the barn was provably used as a hay barn for dairy cattle. The other structure on the farm—20' wide by 30' long—was probably built as a **stable**, although at one time it may also have been used for poultry.

L5: Lopez Village. EuroAmerican settlement at Lopez Village began with the establishment of a store/trading post by Hiram E. Hutchinson in the 1850s at the site of a Coast Salish village named *Sxolect*. In 1879, Hutchinson obtained a homestead of 160 acres, encompassing the main area of the current village. Farmers such as Oscar and Bert Weeks cleared the land and established orchards as part of the islands' burgeoning fruit industry. Later, they established more diversified farms, such as dairy operations.

L6: Lopez Island Historical Museum (28 Washburn Place). The Lopez Historical Museum is a great source of exhibits and historical information on farming on Lopez Island and on the San Juan Islands in general. It has an extensive archive of photographs as well as files of Lopez family history.

L7: Water Towers. Prosperous farmers constructed towers to store water in a tank high above the ground, providing pressure through gravity. Water towers, also called tank houses, were built adjacent to a well, with an associated structure that housed the gasoline pump used to raise the water to the level of the tank. There are three water towers in Lopez Village, all associated with the Weeks family. Parents Lyman and Irene Weeks moved from California to Lopez Island in 1874. Irene was the sister of Hiram Hutchinson, who settled and homesteaded the land that became Lopez Village; she was the island's first postmistress. The three towers were built by sons Oscar (ca. 1914), Bert (ca. 1914), and Edson (1915–1916). All three structures are three stories tall. The top story of each tower held a round tank constructed of wooden staves with wood or metal hoops. All are 14' square at the base, 30' tall, and constructed of tapering walls framed with 2x10s and clad in shingles. In some the ground floor used to hold the milk room housing a separator. The middle or second floor was simply used as access to the tank above; all have been modified for storage and other uses.

L8: Bert Weeks Barn. This structure, demolished since this book was written, used to be a substantial dairy barn. Constructed of a wood frame with milled members, it measured 38' wide by 44' long and is 28' 8" high. The gambrel roof covered a central haymow that was filled by means of a hay track-and-trolley system. Doors in the middle of the sides accessed the mow as well as the horse stalls and cow stanchions that flanked a center aisle on the south side.

L9: Joe Burt Farm. Joseph Burt moved to Lopez in 1902. A carpenter by trade, he worked with his brother John on several

Lopez buildings: the Center School (now Lopez Grange [L14]) and the Mud Bay School, as well as several residences including the Robert Jones House (1906) and his own "House of the Seven Gables" (1908). There are several agricultural structures remaining from his farm located at the rocky edge of the tree line overlooking the surrounding fields, including a barn, granary, milk house, and machine shed. The main 30'-wide-by-60'-long portion of the **barn** under the gambrel roof is two stories, with a 10'-high ground floor and a 28' 8" high hayloft; the shed to the east has stalls. The structure is milled wood post-and-beam, and there is an ingenious hay elevator that takes bales from the ground level up into the loft and then along the upper part of the loft for placement. The **granary** is a 16'-wide-by-22'-long gable-roofed structure, with shiplap siding and few openings except for a central door—typically "tight" to exclude vermin from the grain stored inside. The **milk house** is one of the larger ones in the islands, with a broad-spreading gable roof; inside, the walls are whitewashed for sanitary reasons. The **machine shed** is a 24'-wide, 90'-long shed-roof structure, composed of two parts: a 40'-long chicken house and a 50'-long machine shed. The ruin of an old 13'-wide-by-15'-long log cabin—possibly a "starter home"—still stands near the road.

L10: Section Line Doglegs (Kjargaard Road). As you proceed south from the Joe Burt Farm, you encounter several doglegs in the road, some very closely spaced. These follow the orthogonal grid of the Public Land Survey System's township-and-range section lines. In relatively flat topography, rather than follow contour lines, farmers chose boundary lines for their roads so that they could split the right-of-way easement (which took away from valuable farmland) with their neighbors.

L11: Chris Jensen Farm (Stonecrest Farm). The aptly named Stonecrest Farm is located on top of a rocky knoll, with fields spread out to the west and east. Originally farmed by Chris Jensen, the property was purchased in 2017 by the Lopez Community Land Trust as part of their Lopez Island Farm Trust initiative. The former owners of the farm have a life estate on the land, while tenants have been hired to farm the 46-acre property.

The farmstead consists of several buildings, including a house, barn, chicken house, milk house, machine shed, root cellar, granary, and smokehouse. The main gable of the **barn** is 30' wide by 40' long, with an 18' 6" shed on each side, forming the broken gable. A timber-frame structure, the center portion was used as a loose haymow, and the sheds were probably used for stanchions and equipment storage. Near the barn is a 28'-wide, 66'-long **machine shed** that was probably built in several sections over the years to house the many farm implements. The **house** was originally a gable-roofed, one-and-a-half-story structure with a large gable dormer looking south; it has been modified several times. Attached to the house is the 8'-wide-by-10'-long **milk house**. Nearby is a combination **root cellar-and-granary**; the "cold room" portion has sawdust in the stud walls for insulation while the granary is very tightly built to prevent access to the grain by vermin. The 12'-square **chicken house** is a simple box construction—vertical planks nailed to a sill and a plate—and probably housed several dozen birds. Finally, the 6'-wide-by-8'-long-by-9'-tall **smokehouse** is further removed from the other buildings to prevent the smoke from affecting the dwelling.

L12: Harry and Lillian Towell Farm (Sunnyfield Farm) (6363 Fisherman Bay Road). Harry Towell was born in England and immigrated to Minnesota in 1882. There he met and married Lillian May Stevens; they had three children, Harry Leo, Charles Percy, and Gertrude May. In 1898 the Towells, along with forty neighbors, hired a Northern Pacific Railroad sleeper car and came to Washington State. The next year they purchased farm property in Lopez's Center Valley. The Towell Farm grew grain crops, peas, and hay, and milked dairy cattle. When the telephone came to Lopez Island, the family ran the long-distance switchboard. Harry was also involved in local politics: in 1910, he ran (and lost) on the Socialist ticket for the state legislature against J. W. Frits; in 1912, again as a Socialist, he ran for County Commissioner from Lopez (Third District) but lost to Thomas G. Blake. The Towells sold their property in the 1920s and moved to Bellingham, where Harry died in 1936. After several subsequent owners, Andre and Elizabeth Entermann began a goat dairy operation on Elizabeth's parents' place, and in November of 2014, the Washington State Department of Agriculture granted them a license to sell raw goat milk, aged cheese, and pasteurized cheese. Sunnyfield Farm sells its products to local stores and restaurants, as well as from a farm stand and at farmers markets in Friday Harbor and Lopez Village.

On the farmstead are a house, barn, chicken coop, milk house, and granary. The broad-gabled **house** is built in the Craftsman Bungalow style with a large gable dormer and porch to the south. What remains of the gable-on-hip-roofed **barn** are two shed-roofed wings in an *L*-shape plan. The two wings measure 20' wide by 79' 6" long north-south and 20' wide by 30' 2" long east-west; both are 17' high. The north-south wing has stanchions, mangers, and a manure gutter; the east-west wing has the remains of stalls—all indications of a dairy operation. The **chicken coop**, which has been extensively remodeled, is 12' wide by 20' long; an inscription in the concrete porch—Dec 20 1916—is probably close to the original date of construction. The **granary** is a narrow (24'), long (40'), gable-roofed structure with sheds. The center 16' section, judging from its tight, vermin-proof construction, was probably used to store the grain, while the other portions could have been used for other storage. The **milk house** is a standard 8'-wide-by-12'-long gable-roofed building with 2'- to 3'-high concrete stem walls.

L13: George Boulton–John and Lena Wilson Farm/Gerhart and Frances Kring Farm. John Henry Wilson, who came to Lopez from Michigan in 1892 when he was 15, married Lena Boulton in 1897 and established an 80-acre dairy farm with his father-in-law, George Boulton. The farmstead includes a house, barn, and blacksmith shop (Wilson was the smithy) on the property. They sold the property in the early 1940s to Gerhart and Frances Kring, who continued to farm it. The **barn**, which is listed on the Washington Heritage Barn Register, has two parts: a 46'-wide-by-70'-long-by-31' 9"-high broken gable-roof section used for hay, grain storage, and horse stalls and a 30'-wide-by-36'-long addition to the east that contains a "modern" milking parlor, complete with concrete floors, manure gutters, and pipe stanchions—indicating the extent of the dairy operations. The **blacksmith shop**, which was moved south from the nearby corner, is two stories: the walls of the ground floor were lined with tools and implements of the trade, while the upper story was often used for dances. The **house**, which is located across the road, is two stories, with four full-height columns that support a porch below and a balustraded balcony above.

L14: Lopez Island Grange #1060 Hall (452 Richardson Road). Originally built around 1900 by Joe Burt as the second Center School, the Lopez Island Grange #1060 Hall is an L-shaped building with a large gable facing and a hipped hall parallel to the road; the Roman-arched doorway clearly indicates the entry. The Grange, or Patrons of Husbandry, was formed as a fraternal organization following the Civil War (1867)—and one of the first to offer full membership to women. It has always been supportive of agriculture and farm families. The Lopez Island Grange was first organized in 1909, reorganized in 1914, and then a new Grange (#1060) was organized in 1935. The Grange acquired the Center School as a Hall in 1944 and a new addition was built in 1965 to enlarge the kitchen.

L15: Ruth Bates/Clarence M. Tucker/Barney and Marguerite McCauley Farm (Case Farm) Barn. Not much is known about the history of this barn. Ruth Bates purchased the land through preemption in 1892, and the house and barn were built around 1910 just north of the property. It is likely that the house and barn were moved here in the 1920s or 1930s, when Clarence M. Tucker, an 1890s mill operator and the postmaster at Argyle on San Juan Island and County Treasurer, used the land to raise fruit. The subsequent owners, Byron Norbert "Barney" Goodrow and Marguerite McCauley, raised fruit and produced dairy products on the farm. The barn consists of a main, gable-roofed area, 24' 6" wide by 48' 5" long by 19' 6" high, that was used as a haymow, and a 16' shed on the south side that forms the broken gable, which was used for milking. There are remains of 14 to 15 stanchions as well as mangers, a manure gutter, and an access ramp for the milch cows—an average-size dairy operation in the islands.

L16: Steinbruck's Place Barn. Built around 1915 as a hay barn, this structure has also been used over the years as storage for pea straw as well as tarred reef nets. The center-entry, saltbox-roofed structure measures 45' wide by 65' long and is 35' high. A distinctive feature of the structure is the use of scissor trusses to support a portion of the roof. It is listed on the Washington Heritage Barn Register.

L17: Richardson. George Stillman Richardson, from Mount Desert Island, Maine, received a patent for his homestead here on November 25, 1879, when he was 33 years old. Although he and his family only stayed a few years, the settlement kept the name Richardson as it became an important shipping port on the south end of Lopez. William Graham moved here with his wife Mary from Iowa and established a post office in 1887; in 1889, he and his stepson Thomas Hodgson built a wharf and then a warehouse in 1890, the same year Graham helped Robert Kindleyside establish a store there. (Other members of the Graham family also moved to Lopez: James Cousins and Elizabeth, wife of Albion K. Ridley.) Soon Richardson was exporting agricultural crops and importing supplies. In the mid-1890s the Salmon Banks fish trap and cannery were established here. After the Salmon Banks Cannery burned down in 1922, Ira Lundy planted loganberries on the fields west of the dock and store. This offered the out-of-work cannery hands employment. Richardson continued to be an important port until 1990, when the general store and warehouse burned down.

L18: Hodgson Barn. Norman Peter ("N. P.") Hodgson moved to Lopez in 1878 with his stepfather William Graham, married Lillie Schmaling in 1897, and became a prominent member of the Richardson community, running the store there after Graham bought it in 1899. This barn may have been one of several located on his farm surrounding Richardson. It is approximately 74' wide and 32' deep, and consists of a 32'-square center section with a loft flanked by 20' sheds. At its peak, it is about 26' high. According to tradition, the center portion was moved from a nearby site prior to 1900; later the two side sheds and a back shed (since removed) were added in the early 1900s. The center portion, which has a concrete foundation and floor, functioned as a granary.

L19: Owen and Eva Higgins Farm Barn. Built by Owen James Higgins in 1938, the barn was used as a dairy barn until 1955, when the San Juan Dairymen's Association stopped picking up cream on Lopez for the creamery in Friday Harbor. Higgins, who was a carpenter and shipwright, built the barn himself. The Dutch gambrel-roofed structure measures 38' 6" wide by 72' 6" long and is 38' high. It displays typical features of a dairy barn: a lower story with a center aisle flanked by stanchions for the dairy cattle and an upper hayloft under the gambrel roof, which allowed for large piles of loose hay. It is listed on the Washington Heritage Barn Register.

L20: Ab and Elizabeth Ridley Farm. The Ab and Elizabeth Ridley Farm which included a barn and milk house, is clearly visible from two viewpoints: driving east on Vista Road, the farm appears spread out along the rise above the lower farmland; and more close-up from Mud Bay Road. Albion ("Ab") K. Ridley was born in Maine and fought in the Civil War. He arrived with his wife Elizabeth Graham on Lopez in 1902, and they mainly ran the Richardson Hotel, so it is not clear who actually farmed this property. The timber-frame **barn** has a center drive with flanking haymows under the gable-roofed portion and flanking stalls under the shed that forms the broken gable. It measures 30' wide by 54' long, with a 15' shed; the height is 35' to the ridge. Of particular note are the sills resting on fieldstones. There is also a **milk house** near the road, which is 17' square and features a 2'-deep whitewashed overhang, used to shelter the "milk-run" deliveryman who came to pick up the cream and take it to the creamery in Friday Harbor.

L21: James Cousins Farm. James Cousins and his family homesteaded this place in 1883 at the urging of William Graham, a relative, who had settled at Richardson in 1877. Cousins had a house built on another site on the farm, and then moved it here, eventually adding a barn, machine shed, milk house, smokehouse, and root cellar to the rocky rise of land at the edge of the trees that overlooked the fields below. The timber-frame **barn** has a gable roof with sheds on two sides; the main roof, which measures 30' wide by 60' long, covers a center drive that follows the contour line of the slope. The barn is 35' 3" high, and there is no evidence of a hay rail-and-trolley system. A lower, 18'-wide shed, forming a "broken gable," contains the milking area, with stanchions, mangers, and manure trough; the cows could enter from one side and then exit the other after being milked. The other shed contained stanchions on one side and stalls on the other. The 25'-wide-by-60'-long **machine shed** also follows the contour of rise, so that equipment could be parked off the road leading to and through the barn. The **house**, further to the east, is *L*-shaped with a gable roof and a porch filling the *L*. Further still is the **milk house**, which is 12' wide by 14' long and has a typical 12"-high concrete stem wall. Upslope from this is a 6'-wide-by-10'-long **smokehouse**. Finally, built into the slope to take advantage of the coolness of soil, is the **root cellar**, featuring *V*-notched logs, a poured-concrete floor, and a ventilation shaft to allow for exhaust of warm air.

L22: Peter Nielsen Farm. Peter Niels Nielsen built a modern dairy barn, milk house, and loafing shed here in 1958. The modern **barn** is only one-story high (15' 5") and has a low gable roof over a frame construction on 48"-high concrete stem walls. The main portion has a concrete floor with a drive-in area, with mangers on the sides and stanchions under the portion of the roof that extends to the east. A **milk house**, separated by a walkway, is sheltered under an extension of the roofline. To the south, under an extension of the main roof, is a **loafing shed**. Other farm-related structures are a farmhouse, an old milk house, a shed, and possibly an older barn.

L23: Joe Ender Farm. Joseph Ender had a farm with several structures: a (remodeled) Dutch Revival–style **house** that he moved from nearby Lopez Hill and a barn, milk house, root cellar, and workshop. The **barn** has a gable roof with a shed added along the side. The main portion under the gable measures 30' 8" wide by 74' 4" long; the center-access drive, flanked by haymows, is 31' 2" high. A lower, 20'-deep shed contained stanchions for milking cows, with piping and spigots at each stanchion for vacuum milking. The milk was taken to the adjoining milk house for separation and cooling. The **milk house** is separated from the barn by a short space; it measures a typical 8' wide by 10' long, is sheathed in clapboard, and has a 3'-high concrete stem wall. Between the house and the barn/milk house are the root cellar and the workshop. The **root cellar**, 10' wide by 14' long, is built of vertical wood boards on a concrete stem wall and only has one door for access. The **workshop**, which also has a concrete stem wall, is 20' wide by 64' long.

L24: Joseph and Susie Gallanger Farm (Midnight's Farm). Filtered through the trees as one heads north on Center Road is the Joseph Gallanger Farm—now called Midnight's Farm. Joseph Fargie Gallanger was born in Michigan and came to Lopez Island in the 1890s, marrying 16-year-old Susie May Cochran of Port Stanley in 1896 when he was 30 years old. They bought this property in 1898 and subsequently farmed it as a dairy operation. Surviving buildings from the farmstead include a barn, milk house, and granary. The **barn**, constructed of log posts with milled beams and braces, has a center-entry drive with a shed on one side to form a saltbox roof. The main gable area, which measures 38' wide by 55' long by 33' 7" tall, probably had haymows on both sides; the 18' shed most likely held milking stanchions. A nearby **milk house** was used for separating and cooling the cream; it is a typical 8' wide by 10' long, with shiplap siding. There is also what may have been a **granary** (it has been remodeled, so it is hard to tell); it is 14' wide by 20' long and built with rabbeted horizontal wood siding.

L25: Sidney and Janey Hudson/William and Annie Gallanger Farm (Horse Drawn Farm) (2823 Port Stanley Road). William Walter Gallanger was born in Michigan and came to Lopez with his family in 1891; he married Annie Eliza Bartlett in 1894. In 1921, he, along with his sons Walter and George, bought this property from Janey Hudson, widow of Sidney Hudson. When William purchased nearby property from Henry and Lavinia Erb in 1943, it came with 54 hogs ("including 5 brood sows") and 103 head of cattle, as well as 4 horses ("with harness") and farm machinery such as a tractor, plow, mower, drill, binder, disc, and harrow as well as a Yellow Kid separator, Bowzer feed mill, and DeLaval milker.

The farmstead has several historic agricultural structures: a barn, milk house, and two very modern Harvestore silos. Built ca. 1925–1926 by Walter Gallanger and his son Bill, this **barn** is one of the few remaining on Lopez that is still used for agricultural purposes. The Gallangers designed the structure as a dairy barn for 59 cows—one of the largest operations in the islands. It measures 30' wide by 100' long, with a 20' shed on the east, and is 39' high. The structure exhibits a mix of old and new architectural technology, with a center-drive plan and a prefabricated "bent" system combining round log posts with dimensional lumber beams, plates, girts, purlins, and rafters. Concrete floors and plinths were added as foundational support for the structure, and a (then) state-of-the-art **milk house** appended to the southwest corner. It is listed on the Washington Heritage Barn Register. The two **Harvestore silos**—the only known examples in the islands—indicate the relative scale and importance of the operation. In 1945, the A. O. Smith Company developed the "Harvestore," which featured fiberglass bonded on the inside to the metal container. Each Harvestore is 20' in diameter and 61' high, painted a distinctive blue, and features a system of unloading from the bottom of the silo. Although they performed better than traditional concrete or wood silos, they were also more expensive: in the 1960s, for instance, a Harvestore cost $11,302, twice the $5,435 for a concrete stave silo.

L26: Charles and Lydia Biggs/John C. Ringler/William and Nell Hayton/Henry and Lavinia Erb Farm (Swift Bay Farm) Barn. Swift Bay Farm, at the intersection of Center and Cross Roads, was home to several farmers, including the Biggs, Ringler, Hayton, and Erb families. The principal historic structure is the barn, whose construction date is uncertain (estimated late 1930s); John C. Ringler purchased the land from Charles and Lydia Biggs in 1905, and then sold it to William Hayton in 1936; Hayton and his wife Nell resold it in 1942 to Henry and Lavinia Erb. The roof of this barn is odd: it is almost a Dutch gambrel, because there is a slight lower-pitched roof on the ridge, but the lower slopes stretch down (with a slight alteration) to the first story on either side—and hence designated as simply "Dutch." Built as a dairy barn, the plan features a floor-to-ridge haymow in the center flanked by open stalls on either side, where the stanchions used to be. Oriented north-south, the structure measures 58' 6" wide by 60' long by 28' tall and features a metal hay rail with hay door and hood on the south side.

ORCAS

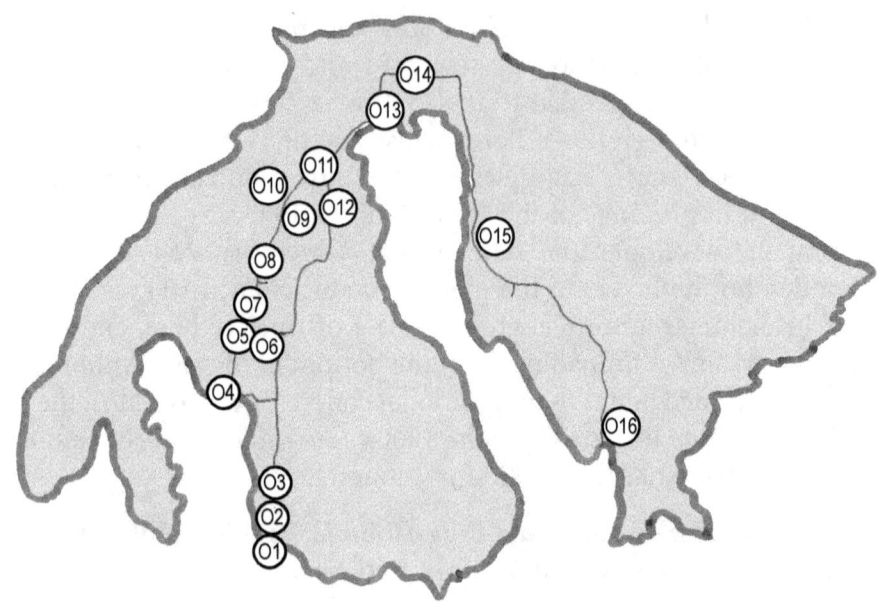

Orcas Island

The island equal in land area to San Juan and second in importance with regard to population and development in this county, is Orcas, which is the leading fruit raising section of the Sound country. The orchards, most of which are still young and not in full bearing, produce large quantities of apples, prunes, pears and cherries. The island is also wonderfully well adapted to stock raising and dairying. It is the best watered of any of the large islands of the archipelago. Its elevated portions afford fine pasturage, while the low lands and valleys produce abundant crops of fruits and miscellaneous farm products of the northwest.

— *The San Juan Islands, an Illustrated Supplement to the San Juan Islander,* 1901

Orcas Island offers a clear example of how geography shaped the agricultural history of the various San Juan Islands. Composed of hilly to mountainous landmasses separated by two major sounds—East and West—the agricultural land is situated in pockets of arable soil such as those around Deer Harbor, Crow Valley, Eastsound, and Olga. Geography also influenced settlement, with several small hamlets or villages situated on good harbors and anchorages. As a result, Orcas Island has four zip codes: Deer Harbor, Eastsound, Olga, and Orcas. There is some evidence of pre-contact Coast Salish farming on Orcas, such as cultivating camas and raising woolly dogs, and there were certainly villages located on the island, but all remaining sites are associated with EuroAmerican settlement through homesteading. The ferry landing is at Orcas Landing, and from there one can travel to several farms on the island.

O1: Orcas Landing. In 1883 W. E. (William Evert) Sutherland, a Canadian, moved to Orcas Island and established a landing, which he called Orcas. Sutherland was appointed postmaster of the post office called "Orcas Island" on August 4, 1884. In 1885, he built a wharf (rebuilt in 1889), warehouse, and storeroom with a public hall overhead, and he supplied water and wood for the multitude of steamers that stopped there. Orcas Landing became a vital spot for shipping agricultural products off-island and importing supplies for farmers. Like many of his neighbors, he also had a fruit farm with an evaporator (the Orcas Fruit Dryer Company).

O2: George W. Gibbs Farm. On the west side of the road as you head away from Orcas Landing was once the farm of George W. Gibbs; a little further along on the east side you will see a road with his name on the sign. George William Gibbs was born in Tewkesbury, England, and immigrated to the United States at the early age of 17, first to New York State, and then to Ann Arbor, Michigan, where he bought his first farm. In the mid-1880s, already well known for his widely published articles on farming in several states and territories, Gibbs chose Orcas Island for his next farm. At first, Gibbs cultivated fruits and nuts on 121.45 acres in Warm Valley near Orcas Landing, leased from the San Juan County Commissioners. Gibbs worked with neighbor George Meyers planting and cultivating orchards. They were two of the first farmers to abandon Italian prune plum trees for apples and pears. After experimenting with hyacinths and noting their proliferation, in 1892 Gibbs ordered $5 worth of hyacinth, tulip, narcissus, crocus, and lily bulbs. By 1898, he had attracted the attention of some Dutch bulb growers, who came to inspect his acreage on Orcas and were favorably impressed. Upon submitting bulbs to the Trans Mississippi and International Exposition of that year in Omaha, Nebraska, Gibbs received a silver medal for his varieties of narcissus, iris, hyacinth, crocus, and tulip, with comments on the particularly fine quality of his Madonna lily. The next year, Gibbs moved to the mainland, where he pursued his bulb growing. Eventually, tulip growing was adopted in the Bellingham area and then, after a particularly hard freeze in 1929, tulip growers moved to the Skagit Valley.

O3: George Meyers Farm Barn. Farther along, Warm Valley opens up and you can see a low-slung barn on the right with two distinctive roof vents: this unique structure was originally used for fruit storage, and its form is signal to its purpose. Builder and carpenter F. J. Reddig designed and built many of the fruit evaporators (dryers) during the boom time of apple and pear production on Orcas Island; he constructed this barn in 1901 as fruit storage for George Meyers. Meyers, along with his neighbor George W. Gibbs, grew several varieties of apples and pears, and even shipped his crop as far as New York City to earn what were then splendid profits. The structure was designed to keep fruit cool on the ground floor by venting warm air through the roof. Oriented north-south, the gable-roofed section, which is one story with an attic, measures 24' wide by 48' long and is 18' 6" high. There is a 12'-wide shed on the west side, with a dormer that allows for access to the center drive and shelter from the elements. Two tall ventilators on the roof drew warm air out of the building, and the frame walls and ceiling are packed with sawdust to help insulate the fruit-storage space on the ground floor.

O4: Crow Valley. Crow Valley drains from the surrounding slopes of Turtleback Mountain down to West Sound, where there was once a wharf, warehouse, and general store. The valley has some of the richest farmland on Orcas, and the farms there boasted production of fruit (apples, cherries, and pears, as well as prune plums), dairy products, and hay. Crow Valley Road runs along the foot of Turtleback Mountain through the valley, revealing many prosperous farms.

O5: William Miller/Henry Mathesius Farm. At the northeast corner of Crow Valley Road and Nordstroms Lane is a complex of several farm buildings, including a large, gambrel-roofed dairy barn, both small and large milk houses, pig barn, machine shed, and farmhouse. The land was originally preempted in 1879 by William Miller, who planted a prune plum orchard and may have built a prune dryer. In 1929, the Mathesius family bought the place and Henry Mathesius established a dairy farm there. The **dairy barn**, built by Ray Pineo in 1933, has concrete floors and stem walls, with a wood-frame lower story and wood-truss hayloft. Interesting features include wooden stanchions, poured concrete manure gutters, and three metal ventilators on top of the gambrel roof. Oriented east-west, the structure measures 42' wide by 64' long, with an overall height of 44' (9' 6" on the lower floor; 34' 6" for the loft). The old **milk house** is standard size—8' by 10'—and sided in "German" shiplap (with a curved groove). A newer **milk house** has been added onto and converted into a guest cottage. The farmstead also features a 30'-by-40' **pig barn** and a 20'-by-74' **machine shed**.

O6: Nordstrom Farm. Settled by Andrew Nordstrom in 1901, this quarter-section farm originally had pear and Italian prune plum orchards and prune-drying operations as part of the early Orcas Island fruit industry. The Dutch-roofed, 40'-wide-by-60'-long-by-30'-high **barn** was used primarily for hay storage, with stanchions for milking dairy cattle. It is listed on the Washington Heritage Barn Register. A **log structure** with a frame addition was the first residence on the farm; it was later used for fruit processing and packing. The second **farmhouse**, which is frame construction, has been renovated.

O7: Bert Chalmers Farm Barn. Located on the eastern hillside of Turtleback Mountain, the barn on the old Chalmers Farm is sited on the slope to take advantage of different levels for different farm functions. The main gable structure is a five-bay timber frame, with the shed forming the saltbox added downslope to house a milking shed below and a loft above. The structure measures 26' wide by 50' long and is 27' tall; the saltbox shed extends 17' downslope to the east, and there is another 18' open shed on the north gable. The haymow can be accessed at floor level, while the cows enter the milking shed at a level downslope. The 160-acre farm was purchased in 1900 by Welshman Albert "Bert" Earnest Murray Chalmers, who immigrated to Orcas Island with his father Alexander and older brother Estyn Murray in 1891. (Estyn had a nearby farm where he raised red poll cattle.) Bert sold the farm and moved off-island in the late 1910s.

O8: Turtleback Inn Barn. On the property that now features Turtleback Inn, there is a barn that was built as part of a farm originally homesteaded by Peter Frechette in 1894. It is a simple gable-roofed structure, 24' wide by 40' 8" long by 25' 6" high, that was built into the slope on a north-south axis. Sheds have been added upslope on the west (6' wide) and downslope to the east (20' wide). Although there are log posts in the main section, most of the structure is milled lumber, indicating newer construction.

O9: Orcas Island Winery (2371 Crow Valley Road). Established in 2011, Orcas Island Winery is one of four vineyards in the islands. It has approximately two to three acres planted and a tasting room where the owners currently offer nine varietals.

O10: William Hambly Farm and Apple House. William B. Hambly settled on this place in 1882 and applied for a homestead, which he was granted in 1889. Hambly's homestead, which eventually grew to include 600 apple trees and 1,000 prune trees, once included a house, barn, chicken house, wood shed, and smokehouse. Hambly also had a prune dryer operating on the place in the 1890s. Built in 1900, the two-story apple house was designed to keep apples cool before being shipped. The structure measures 24' wide by 30' 4" long, with the basement height at 8' 2", and the superstructure height of 13' 4". It is built into a slope so that one can access the basement from the lower ground and the upper story from up-slope. The lower, 20"-thick stone-masonry portion has vents near the ground that introduce cool air that then circulates upward through the vent in the wood-frame roof on the floor above. To help insulate the basement, the spaces between the joists of the ceiling were filled with sand. The upper floor was probably used for fruit-box construction and storage.

O11: Auld Farm. At what locals call "Fowlers Corner," where Crow Valley Road meets Orcas Road, is a long, low structure that is raised on piers above the ground sloping beneath it. Constructed by Robert Auld on land purchased in 1906, the structure was part of a larger farm complex that included a barn and workshop. (Auld's daughter Grace married Harry Fowler, who constructed nearby Fowler's Pond.) The Aulds raised sheep and cattle and hayed the fields for their livestock. Several articles in the *San Juan*

Islander noted that Robert Auld was involved in the breeding of Ayrshire dairy cattle, mainly from Willowmor Farms of Redmond, Washington. The **poultry house**, which measures 20' wide by 70' long—indicating a large operation with hundreds of birds—features large windows on the south side and smaller windows on the north side where the poultry cages were. The saltbox roof allowed for a taller interior hallway to access the shorter caged areas.

O12: Senior Class Shed. On Orcas Road across from Fowler's Pond is an old shed structure that was once part of a prune dryer. Prune dryers were usually located near a stream or other source of water for rinsing the fruit, as is this one. The shed probably contained the stove used for heating a larger, square, two-story structure with a pyramidal roof with a vent on top—the chamber used for drying the fruit. For years, it had been decorated by the current Orcas High School senior class; recently, because of the fragility of the structure, the owners built a new shed, which high school seniors now use for decoration.

O13: Eastsound. Agriculture was the foremost economy of Orcas Island, and Eastsound used to be surrounded by farms and orchards; historic photographs indicate fruit trees—principally Italian prune plums—throughout the area. This was such an important business that one of the main streets is called Prune Alley. The Orcas Island Historic Museum, located at 181 North Beach Road, consists of six log cabins that were moved from their homesteads and connected to form a series of exhibition spaces. The museum features exhibits on the history of the island, including agriculture, with the fruit industry prominently featured.

O14: Jessie Waldrip Farm (706 Mount Baker Road). On the north side of Mount Baker Road is a tall, Gothic-roofed **barn**, which is unique: it is the only known extant Gothic-roofed barn in the San Juan Islands. Jessie Murphy Waldrip purchased this property in 1912 and hired W. R. B. Wilcox, an architect who practiced in Seattle from 1907–1922 to design a dairy barn, which was built in 1918. Based on a series of letters between Waldrip and Wilcox, the architect used standard plans from the Louden Company, featuring a huge haymow and loft under the Gothic roof and stanchions for milking the cows on the ground floor. (Wilcox's bill for his services came to a total of $23.80—or only $436.49 adjusted for today's inflation.) The two-story barn measures 44' by 60', with stanchions, stalls, and work spaces on the ground floor and a 31' 6" tall hayloft, which features a hay hood, wooden hay rail, steel trolley, and cupola ventilator. The correspondence between architect and owner discusses several aspects of the design and construction: loft capacity, arrangement of stalls, and shingling vs. board-and-batten siding, as well as the design of other farm buildings, including a silo and root house. Mount Baker Farm is also known for the later addition of a standard-gauge railroad system, replete with tracks, water tower, train sheds, and other rail-related buildings. The farm currently hosts seasonal camping.

O15: James Francis and Annie Tulloch Farm. James Francis Tulloch came to Orcas Island in 1875, married Nancy Anne "Annie" Brown, and raised nine children. In his eponymous *The Diary of James Francis Tulloch, 1875–1910* (a memoir compiled and edited by Gordon Keith in 1978), Tulloch offers detailed descriptions of farming on Orcas, including his prominent role in the nascent fruit industry. His farm was located at this spot, although the current barn, which looks old, is not the original.

O16: Olga Strawberry Barreling Plant (intersection of Olga and Point Lawrence Roads). The Olga Strawberry Barreling Plant, located in the hamlet of Olga, was constructed in 1938 and stands as a reminder of the once-flourishing strawberry industry on the east side of Orcas Island. Built in cooperation between the Orcas Island Berry Growers Association and the National Fruit Canning Company, the plant provided jobs for locals throughout the Depression and into World War II.

The Olga Strawberry Barreling Plant is a simple wood frame building measuring 30' wide by 80' long, with a 10' dock on the east end. The floor structure is hefty, having to support the barrels filled with 425 pounds of strawberries. The two long walls under the gable roof have 3'-by-3' wooden frame windows with six lites each, arranged in clusters of three to afford the best lighting for the operations inside. The clear-span wooden rafters and cross ties of the roof structure are exposed, as are the 6x9 posts and 2x4 studs of the walls, clearly indicating the utilitarian nature of structure.

Strawberries were picked, hulled, and packed into flat crates of 12 boxes each in the field and then brought to the plant in trucks. After registering and weighing the berries on a floor scale just inside double doors on the west side of the building, workers poured them into a tank of running water and then sprayed them clean on a chain-metal conveyor belt before grading them on a rack. Women hand-sorted out the green and non-hulled fruit, and then another belt sorted the berries into groups of small, medium, and large berries. The fruit fell from a trough into continuously jolted barrels, which packed them solidly; each 425-pound barrel had 318 pounds of berries layered alternately with 107 pounds of sugar. The barrels—about 17 per day—were loaded from the east dock onto trucks, shipped by ferry to the mainland, and driven to Everett for freezing. In addition, diggers brought the year-old, certified Marshall strawberry plants to the building where they were cleaned, trimmed, and packaged in bundles of 25 for shipping.

After processing ended in 1943, the building was used for storage until it was sold in 1978 and converted into a restaurant, the Chambered Nautilus. At that time the owners resided in the structure, remodeled the interior, and added a loft with dormer windows on the west end. In 2004, the Olga Strawberry Council was formed to manage the building as an artist's cooperative and the Café Olga, which Marcy Lund established in 1981 and ran for 22 years; the same year, it was listed on the Washington Heritage Register. In 2006 the Council established the Historic Preservation and Conservation Easement with the San Juan County Land Bank to preserve the unique character of the structure.

On the night of Friday, July 19, 2013, fire destroyed the east deck and restrooms, as well as causing smoke damage throughout the eastern half of the building. A successful response by local firefighters saved the entire structure from being razed. Relying upon the engaged support of the local Olga and overall Orcas Island community, the Olga Strawberry Council has restored the historic character of the building while upgrading it to modern building and health codes. It is currently home to the Orcas Island Artworks Gallery, James Hardman Gallery, and the Catkin Café.

SAN JUAN

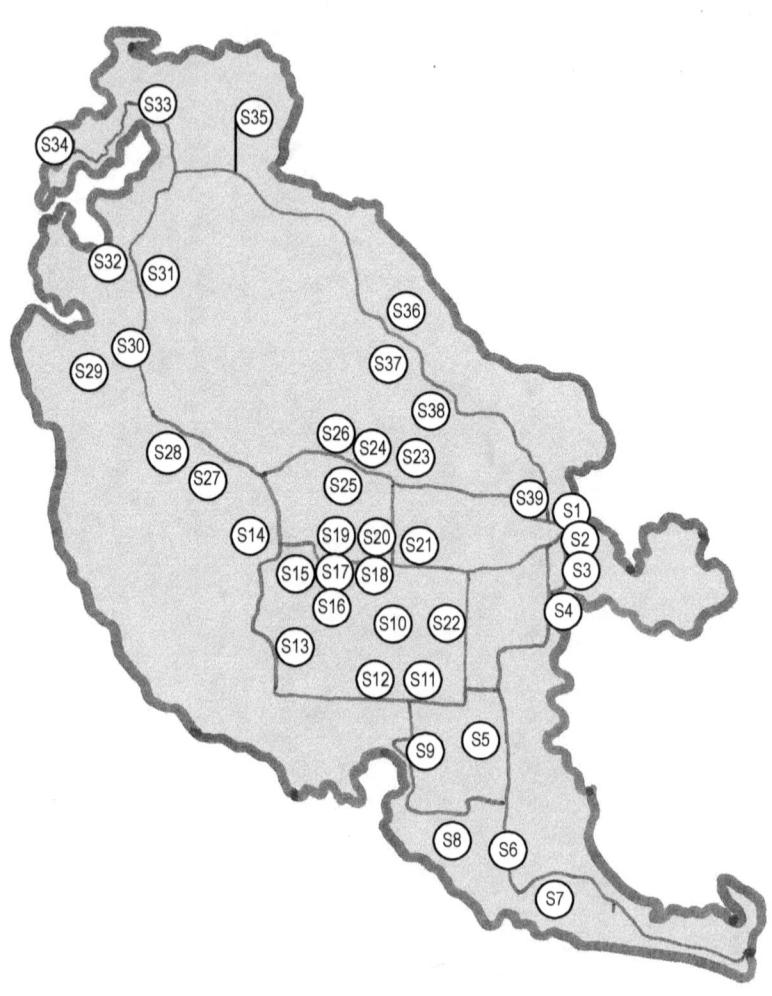

San Juan Island

The island after which this county takes its name and the one recognized as having the largest population and most wealth, is San Juan. The main dependence is agriculture, sheep, and stock raising.... In the San Juan valley, reaching from Friday Harbor to Cattle Point, from Friday Harbor west and north, and surrounding Roche Harbor, are found [the] most desirable farming communities and sheep and cattle ranges, the low lands being adapted to cultivation and the higher ground to grazing.

—*The San Juan Islands,
an Illustrated Supplement to the San Juan Islander,* 1901

San Juan Island has a mix of rich bottomland, grass prairies, and forested highlands that makes for a diverse agricultural environment. EuroAmerican farming first occurred on San Juan, when the Hudson's Bay Company's Belle Vue Sheep Farm established a series of pastures on land used and kept clear by the Coast Salish. During the homesteading period, the island's rich regions of agricultural land—primarily San Juan Valley—provided the basis for growth in island farming, and that legacy is still apparent to the person traveling its roads.

S1: Friday Harbor. During the 1850s the Hudson's Bay Company established a sheep station near this harbor and put a Kanaka shepherd named Friday in charge. Friday lived in a log house located on the present University of Washington Friday Harbor Labs property. The place was eventually named "Friday's Harbor"—presumably after the person Friday. As a consequence of the Pig War—a conflict over possession of the islands—in 1860 British and American troops settled on a joint occupation of San Juan Island. The British, in looking for a suitable place for their encampment, visited several sites, including "Friday Bay." (The British eventually decided upon Garrison Bay as a site for their camp.) Soon after the 1872 settlement of the boundary dispute, Friday Harbor was founded as the seat of newly formed San Juan County in 1873.

Downtown Friday Harbor, and Front Street in particular, was the epicenter of agricultural trade on San Juan Island. Because the islands were served by a fleet of individually owned and operated steamers colloquially called the Mosquito Feet, ports such as Friday Harbor were the center for exporting farm crops and importing agricultural machinery and supplies. Sheep were driven down Spring Street and penned on the docks, awaiting shipment to Puget Sound ports; grain and produce were loaded onto ships, and supplies were unloaded to mercantile warehouses such as the Sweeney Mercantile Company (1870s), which was succeeded by the Bugge Trading Company (1906) and the San Juan Agricultural Company (1909–1961).

Several large structures served the island's agricultural community on this waterfront. The Friday Harbor Creamery, located between present-day Downriggers and the offices of the Port of Friday Harbor, was established in 1901, and churned thousands of pounds of cream into butter every day. (In 1927, the creamery, then run by the San Juan Dairymen's Association, was moved several blocks up Spring Street to a newly constructed plant.) Located nearby on the shoreline (and indeed extending on pilings over the water) was the San Juan Islands Cannery Company, where 4,000 cases per day of peas harvested in San Juan Valley were packed into cans labeled "Saltair Peas." In the 1940s the floor of the building gave way and hundreds of cases of peas sank to the bottom of the harbor. According to locals, the sodden cases were brailed to the surface and the cans were washed, relabeled, and repacked—all with labor from local schoolchildren. Later, George Jeffers used the old salmon cannery of the Friday Harbor Packing Company, located where Cannery Landing is next to the ferry dock, to can peas in the late 1950s; in 1960, he built the Friday Harbor Canning Company, a freezer operation, and froze peas that were raised in the valley until the late 1960s.

It was also on the Friday Harbor waterfront that the first San Juan County Fair was held in 1906. Organizers used several buildings of the Friday Harbor Packing Company as exhibition halls, and in succeeding years the annual fairs were held in other locations downtown, until the fairgrounds were constructed in the 1920s.

Beginning in the 1890s, Friday Harbor's downtown core was surrounded by orchards, particularly Italian plums for prune making. On a map drawn by the US Coast and Geodetic Survey in 1895, hundreds of dots, each representing a fruit tree, surround the built-up portion of Spring, West, First, and Second Streets. G. B. (Granville Baber) Driggs, who had a mercantile establishment on Spring Street, planted extensive orchards (as well as strawberries) on his fields east of Argyle Wagon Road, and built a dryer to convert his plums to prunes.

The **San Juan Island Grange #966 Hall**, located at 152 First Street North, was originally built as a Methodist Episcopal church in the 1880s. The Women's Club bought the property in the 1920s, and eventually sold it to the Grange in 1975, with the provision that it continue being used as a public hall. San Juan Island Grange #966 was chartered in 1931 and has served the island agricultural community ever since. In 1937, local members bought property on Spring Street and built a cooperative farm store with volunteer labor and materials. It offered freezer space, truck scales, and a gasoline pump. In the 1940s the farm store was closed and leased as a gasoline station and repair shop. San Juan Island Grange #966, currently including over 100 members, meets at the Hall the first Wednesday evening of the month, and also hosts public programs. The Grange Hall is a popular venue for islanders' celebrations and community events.

The **San Juan Historical Museum** (405 Price Street), which houses an exhibit on agriculture in its Museum of History and Industry Building, is located on one of Friday Harbor's historic farms: the **James F. King Homestead**. James Franklin King was born in Yamhill County, Oregon; in 1877 his family moved to the San Juan Islands. On Christmas Eve 1880, he married 16-year-old Adeline Verrier of West Sound, Orcas, and the couple eventually had three sons and three daughters. King arrived in Friday Harbor in 1887 and filed for his homestead in 1890. On what eventually became a 445-acre farm, King cultivated 80 acres, raised sheep and cattle (principally dairy), and planted an orchard of some 300 apple and pear trees.

King's **house**, which was built in 1897 after his log cabin burned down, consists of a one-and-a-half-story section with a one-story wing to the north. At one time, the house was surrounded by other farm-related structures, such as a wagon house, barn, dovecote, and granary. Over the years, some of these buildings were demolished or burnt (the original barn, for instance, was demolished in 1937 and a new, Gothic-roofed dairy barn was built in its place, only to be destroyed by fire two years later), while new ones were constructed, reflecting the changing nature of the farming operations. Besides the farmhouse, the **root cellar** (1903, built by mason Charles McNallie) is the oldest structure on the grounds. It is constructed of stones set in lime mortar and plastered (pargetted) with a lime mixture heavy on sand and scored in a *trompe d'oeil* manner to resemble a standard ashlar (rectangular block) wall. Historic photographs of the farm indicate that the root cellar once had a second story constructed of frame lumber, suggesting that it was used for fruit storage in the (lower) stone part and fruit-box production and storage as well as ventilation in the upper story. Later, the **carriage house**, with its concrete stem walls and floor and its dimensional wood-frame construction, was added on. The **milk house** is a typically small (8' by 10'), one-story, gable-roofed building, constructed of a wood frame on a concrete pad and stem-wall foundation. It housed a separator—you can see the bolts where it was attached to the floor—which separated the milk into cream and skim milk.

The San Juan Historical Museum grounds also have outdoor displays of farm equipment: an Advance-Rumely thresher, a millstone from the mill at Argyle, and several cultivating tools. Meinrad Rumely of Laporte, Indiana, began building threshers in 1852 and continued in the business until his death in 1904. Reorganized in 1915, the firm he founded became known as the Advance-Rumely Thresher Company. The stationary thresher, driven by a long belt attached to a steam engine, pushed the grain and straw over vibrating racks. The grain, smaller and heavier, fell through a sieve to the bottom and into sacks, while the straw continued to the back of the machine where it was blown into a stack.

In 1886 a group of men from Port Townsend, Washington, established a mill at Argyle, a shallow water port on Griffin Bay south of Friday Harbor. This three-story mill was powered by steam and ground the various grains—barley, oats, and wheat—on millstones like the one displayed on the museum grounds. Although a kiln was added for drying prunes, the mill closed in 1909 due to lack of business. The town of Argyle, once thought to become the main settlement on San Juan Island, closed its post office in 1912.

Farmers used a variety of horse- or tractor-drawn implements to cultivate the soil and harvest crops. A harvester cut peas and swept them into a row for gathering. A cultivator with spring-tensioned, shovel teeth tilled the soil, keeping it free of weeds. A two-bottom sulky (riding plow), with two moldboards, helped break and turn the sod. A spring-tooth rake was used to rake the hay.

S2: Argyle Wagon Road. If you head out of Friday Harbor on Argyle Avenue, you will be following the route of the old Argyle Wagon Road, which was built by volunteers in 1890 to connect the town with the settlement of Argyle, which was the site of the only gristmill on the island. The whole area to the left past Nichols Street used to be prune plum orchards, and there are a few remaining fruit trees in the area. In addition to town businessmen, several farmers also built "town" or "retirement" houses along Argyle.

S3: San Juan County Fairgrounds (846 Argyle Avenue). Starting in 1906, the San Juan County Fair was held at a variety of locations in Friday Harbor. In 1923, several citizens incorporated a nonprofit San Juan Fair with capital stock of $5,000 and sold 1,000 shares at $5 each to raise money for constructing the fairgrounds on the existing site, at the edge of town. Several buildings were originally constructed: the Pioneer Cabin, a main exhibition building, and possibly poultry and stock buildings. The County Agent was requested to engage the services of a University of Washington landscape architect to design the fairgrounds. The first fair at the new site was held in October 1924; newspaper coverage mentions fruit, floral, and stock displays. The Fair Committee decided to construct a "Pioneer's Building" to commemorate the role of early families in the settlement of the islands, and on July 31 they issued an appeal for a "peeled log" from each pioneer family. The resulting **Pioneer Cabin** was dedicated during the first fair on October 27. An older portion of the **horse barn**, which may date to the original layout of the fairgrounds, remains from a long structure that incorporated several additions.

S4: Oscar Peterson Farm. At the corner of Argyle Avenue and Cattle Point Road is the Oscar Peterson Farm, which today consists of a farmhouse, chicken house, and barn. Oscar Peterson inherited the property from his father and mother, Charles Victor and Carrie Mathesius Peterson; Charles had been deeded the property by his father P. E. Peterson, a prominent San Juan Valley farmer. The **farmhouse** is a single story under a hipped roof and has two bedrooms in addition to a living room and kitchen. The single-story **chicken house**, which measures 24' deep and 50' long, is typical of its kind: a saltbox roof allows for a person-height hallway under the higher section while the lower portion contains the pens for birds. Two stacks in the roof ventilate the accumulated odors. Oriented east-west, the **barn** has a gable-roofed haymow that measures 24' 6" wide by 32' 6" long by 26' 6" high; one 20'-wide shed on the south forms the saltbox roof and another on the west forms a hip off the gable end. The mow has a barn door and steel rail-and-trolley hay system. The shed on the south, illuminated by a bank of windows, had stanchions with entries on the west and east; the other shed may have been used for stalls.

S5: John Lawson Farm. Originally homesteaded by the Boyce family and then farmed by the Wades, this farm eventually became the property of John Lawson. The farm complex sits on a rise overlooking the surrounding bottomlands. The original farmhouse has been moved. In addition to a barn there is a milk house, granary, and machine shed. Oriented east-west, the main haymow portion of the **barn** is 24' wide by 40' long and is 26' 6" high; a 20'-wide milking shed extends along the south side and wraps around the corner to run along the west side. The hay door is on the east side, accessing a wood-rail and steel-trolley hay system. The milk shed features a vacuum milking system; the cows entered the shed via a ramp and door on the southwest side and were milked by machine in their stanchions. The gable-roofed **milk house** abuts the southeast corner of the barn. The **granary** is built into a slope and has a low concrete basement that may have been used as a root cellar. It measures 11' by 17', is roofed with metal and has tight, horizontal wood siding with only doors as openings. The 30'-by-64' **machine shed,** with two open and two closed bays, lies along the embankment near the driveway so that farm machinery can be easily driven in. It is a very utilitarian structure, consisting of wood piers and posts, a pole-truss roof structure, and vertical wood-board siding with no battens.

S6: Christopher Rosler and Robert Frazer Farms. As you are heading to American Camp on Cattle Point Road, you pass between two historic farms: Rosler's to the east and Frazer's to the west. Both were located on the infamous First Prairie where, on June 15, 1859, Lyman Cutlar's shooting of a Hudson's Bay Company pig rooting in his potato patch precipitated the eponymous Pig War.

Heading south, you first encounter the homestead of **Christopher Rosler**, who mustered out of the US Army at American Camp, married Anna Pike, a Tsimshian from Fort Simpson, British Columbia, and received his homestead patent in 1877, at the age of 37. For his farmstead, he chose a prominent knoll with visual command of his surrounding fields—strategic during that time of raids by northern tribes—and built a **barn and root house** with a masonry first floor and log second floor. You can see this building, the **farmhouse**, and a **granary**, from the road.

Further south and to the west, **Robert Frazer**, who took over his farmstead from Cutlar and tried to get the title to his land through the British authorities, claimed a homestead here in 1883. He and his brother William mostly farmed across the road on fields next to Rosler's. The farmstead consists of a house, granary, sheep barn, and several outbuildings. The **house** is L-shaped, with a gable and porch facing the road. The **granary** is 14' wide by 24' long, constructed of solid shiplap siding and has a single door in the center of the façade facing the access road. The **barn**, which is tucked out of sight further up the driveway, was specifically designed for sheep, with its distinct plan of a central haymow surrounded on three sides by sheds. Oriented east-west, the central, gable-roofed, 26'-high haymow is 19' wide by 24' long, with 16' 6" sheds to the north and south and a 16' shed to the east. Milled-lumber posts and beams support the central structure as well as the side walls and roof. This plan allowed for the sheep to be penned in the sheds surrounding the central haymow.

S7: Belle Vue Sheep Farm/American Camp (4668 Cattle Point Road). From the San Juan Island National Historical Park Visitors Center at American Camp, take a walk to the Redoubt, where you can enjoy the commanding view of Griffin Bay, Lopez Island, and Mount Baker to the east and the Salmon Banks, Haro Strait, Straits of Juan de Fuca, and the Olympic Mountains and Mount Rainier to the south. Looking directly east to Mount Finlayson near the tip of the island, one can see a series of shelves, which represent the former shorelines and beaches when the islands underwent a series of rebounds from the immense weight of ice during the late Pleistocene Era glaciations. You can also see huge boulders—called erratics—that were left behind by the receding glaciers; these were left by farmers because they couldn't be moved, and often became the beginnings of rock piles formed from clearing the fields of stone.

This site is the birthplace of EuroAmerican agriculture in the San Juan Islands when, in December 1853, the Hudson's Bay Company established Belle Vue Sheep Farm. The company built a cluster of buildings (marked by a flag pole to the south), called "The Establishment," which consisted of several houses, at least two barns, a granary, sheep shed, henhouse, and potato pits. The company grazed its sheep on Home Prairie, a camas-collecting grounds for neighboring tribes. The farm also had two fields nearby—called "Big Field" and "Little Field"—of oats, peas, potatoes, and other crops.

In 1859, in response to the Pig Incident, American troops established American Camp on this ridge and constructed the redoubt as a defensive earthwork. When the Americans and the

British decided to jointly occupy the island, the Americans planted several fields, some formerly farmed by the Hudson's Bay Company. After the settlement of the boundary dispute in 1872, the current manager of Belle Vue Sheep Farm, Pomona, Orkney Islands–born Robert Firth, as well as several other Americans, applied for homesteads on former Hudson's Bay Company and American Camp Military Reservation properties. The areas around the campgrounds were eventually planted in orchards and the fields of the prairie plowed and planted in oats and potatoes. Some traces of the farming activity can be seen in the few surviving fruit trees and the piles of stones that were removed from the fields.

S8: Straitsview Farm Barn. Heading back north and turning onto False Bay Drive, after some doglegs you will see to the south the magnificent spreading mass of the barn at Straitsview Farm. Built in 1863 by Peter Lawson, an emigrant from Denmark, this structure is probably the earliest-built extant barn in the islands. It has an English (center drive) plan and gable-roofed haymow with sheds on all four sides. Oriented slightly off the east-west axis, the central space is 30' wide by 60' long by 26' high. The sheds on the west and south sides are 24' wide; the one on the east is 18'; the one on the north is 14'. Given the orientation of the structure and the fact that the prevailing winds come off the straits to the south, the barn was probably positioned to take advantage of the site by funneling the breeze into the center drive to winnow grain on a threshing floor. Among the associated farm buildings are a granary and a machine shed.

S9: False Bay Farm. Passing further along False Bay Drive, with False Bay (the drainage for San Juan Valley) to the west, you will drive through fields with several large horse barns and sheds belonging to False Bay Farm, which was established by Laddia Whittier in 1983. Whittier raised thoroughbred horses.

S10: San Juan Valley (intersection of False Bay Drive and Bailer Hill Road). San Juan Valley forms the core of prime farmland on San Juan Island. The valley is formed by the False Bay Creek watershed, fed by two stream systems—False Bay Creek to the west and San Juan Creek to the east, which drain the watershed from the surrounding hills to False Bay. The False Bay watershed area comprises about 11,697 acres and is defined by Little Mountain and Mount Dallas to the west, Mount Grant and Cady Mountain to the north, and a rise forming a divide from the Griffin Bay watershed to the east. It is crossed by Bailer Hill Road to the south, San Juan Valley Road to the north, Wold Road to the west, and Douglas Road to the east.

When the Hudson's Bay Company built a road connecting its pastures on the island, it encountered a ring of Garry oaks around the upland portions of the valley, so company workers called the area "Oak Valley." As Americans began to settle on the island, the valley was one of the first areas farmed, because it was relatively free of large trees and had good bottomland (soils within the streamshed that retained moisture during dry summers). Some of the earliest homestead patents—dating from the late 1870s, because claimants couldn't apply until after the boundary settlement of 1872—were taken in the valley. The heart of the valley, along San Juan Valley Road, was homesteaded by a mix of Hudson's Bay Company "servants"—Kanakas, French Canadians, and Scots—and Irish immigrants. These farms, initially subsistence-based, formed the basis for larger operations featuring beef and sheep ranching, dairying, and crops such as grains, hops, and peas.

S11: Patrick Beigin Granary. To the northeast is a structure that looks like a barn but is actually a granary. It has a central gable roof with sheds on three sides—technically a gable-on-hip roof—and measures 32' by 45'. The two flanking sheds are open to the south, and at one time were open at their other ends, for driving through. The central portion, under the gable, is a granary, solidly built with inside-out framing (the structural frame is on the outside, thereby offering a smooth surface for the grain bins inside) to deter infestation by vermin (rats and mice). A wagon full of grain could be pulled up alongside the granary and sheltered from rain while the grain was loaded into the bins.

Although it is not clear when it was built, the granary was part of the farm operations on the **Patrick Beigin Homestead** (1888). Beigin established his farm in the valley after mustering out of the

US Army at American Camp and marrying Lucy Morse, a member of the Howcan (Sitka) Tribe, whom he met in Victoria. The farm was once a thriving operation, with a fenced orchard, livestock, and grain crops. None of the original farm buildings except for the granary still exist.

S12: E. P. and Gregory Bailer Farm. Turning west onto Bailer Hill Road, you approach a ridge/hill that was the site of the Bailer Farm. E. P. (Engelbert Phillip) Bailer and his brother Gregory, both born in Hohenzoller, Germany, came to San Juan Island in 1870 and filed for adjacent homesteads at this place (patented in 1883/1884). Although there are no remaining structures on the farm site, the Bailer farmstead was extensive, with a house, barn, hop dryer (oast or hop barn—the only one known in the islands), granary, and many outbuildings, and comprised 280 acres, of which 150 were cultivated. E. P. Bailer boasted that his farm yielded per-acre harvests of one ton of wheat, two and one-fourth tons of oats, two tons of hay, and twelve tons of potatoes. A historic photo looking uphill at this site shows a proud family with their wagon posed before a string of buildings, while another, taken from above the farmstead, encompasses the sweeping view of the valley with a cluster of farm buildings in the foreground.

S13: Pelindaba Lavender Farm (33 Wold Road). Founded in 1998, Pelindaba was established as means of saving historic agricultural land through the cultivation of a low-irrigation, low-fertilized crop: lavender. (Pelindaba is a Zulu word roughly translated as "place of great gatherings," harkening back to founder and owner Stephen Robins' South African roots.) Today you can visit the farm of some 25,000 plants (including 50 different cultivars). A gift shop and nursery (in the gatehouse, a remodeled farmhouse moved from nearby) feature the many value-added products made from the lavender plants' flowers, buds, and essential oils.

S14: Sandwith/Boyce Farm (Animal Inn) and Joe Sandwith Barn. Both these farms formerly belonged to the Sandwith family—Joseph J. and Ellen, who had bought the property from Thomas and Katherine Roberts in 1901. Originally, the land to the east was homesteaded in 1883 by Joe Friday, whose father Peter was the "Friday" that the town was named after.

Sandwith/Boyce Farm (Animal Inn) is a farmstead that contains several buildings: a house, barn, milking parlor, granary, and chicken coop. The **house**, built by Sandwith, is the first building you encounter: a low, gable-roofed, one-and-a-half story Craftsman Bungalow with recent additions connecting it with a garage. The **barn**, built by Merle Boyce after he and his wife Christina bought the property in 1948, has a remarkable spread: 60' wide by 62' long, with a height of 25'. It is noteworthy because there is no indication of a hay door or rail-and-trolley system. The structure has log posts at 10' on center, with milled-lumber braces, beams, purlins, and roof joists. The **milking parlor**, also built by Boyce in the early 1950s, is one of very few in the islands: a large building, separate from the barn, wholly devoted to milking cows. It has been extensively remodeled into kennels. The **granary**, which has also been extensively remodeled, was originally a simple gable structure with tight, vermin-proof construction and a single entry. The **chicken coop** appears to be of more recent construction.

Across Boyce Road is the **Joe Sandwith Barn**, a simple gable-roofed haymow with a shed for dairying added on one side to form a saltbox. Another open shed, for calf pens, was added on the other side. The 21' 2"-high main gable is actually square in plan—22' by 22'—with the milking shed (complete with six stanchions and a manure gutter) extending 15' 8" to the east and the calving shed 14' to the west.

S15: King Sisters Farm. Located at the dogleg in San Juan Valley Road was the farm of the King sisters. All that remains is a granary that is currently used for hay storage. It is 18' by 24' by 14', and is a characteristically tight structure, with only one door, to discourage vermin such as mice and rats.

S16: The Heart of San Juan Valley. As you come out of the dogleg in San Juan Valley Road, you approach the heart of the valley, rich in its history of French Canadian, Kanaka-Salish, and Irish homesteaders. These men and women established farms in the rich bottomland of the valley, cultivating fields for crops and clearing and seeding pastureland for livestock grazing. This pattern can still be seen in the farmsteads you encounter along San Juan Valley Road: agricultural buildings clustered on higher, rocky ground with fields spreading out below them.

S17: Wooden Shoe Farm. As you come out of the dogleg in San Juan Valley Road and look south, you will see Zylstra Lake and a complex of farm buildings that was once Wooden Shoe Farm. In 1960 Fred Zylstra assembled the farmland from at least three homesteads to form a modern cattle-breeding ranch specializing in purebred polled Herefords. Zylstra developed the farmstead from existing structures such as the house, a barn, and a granary, with a building campaign in the early 1960s of a workshop/garage, poultry house, granary, power house, milking station, several loafing sheds, a machine shed, and a new barn.

At the time Fred Zylstra built **Zylstra Lake** (1963) for irrigation and wildlife conservation, it was considered the largest privately owned, man-made lake in Washington State. The **house** has been largely remodeled. The **workshop/garage**, built in the early 1960s, is 30' wide by 60' long by 10' high. A 16'-wide-by-40'-long **poultry house** is sided in board and batten, which are painted red. An old (1900s) 18'-wide-by-28'-long **granary** is sided in the same materials and color, as is the neighboring **power house**, which measures 9' wide by 13' long. Across the driveway is the **new granary**, a 24'-by-24', two-story structure built into the slope; grain could be delivered to the top story and milled to bins below, where it could be loaded into trucks. A newer building, the **milking parlor**, also contained the farm office. There are two **loafing sheds**—open frame structures with feeding troughs running along the center—located near the pastures. Between the old and the new barn stands a **machine shed** that is enclosed on three sides; this is a simple structure used to shelter the various farm implements.

The **old barn** may have been built by John P. Doyle in the 1890s as part of his farm operations. It was used for hay storage and features a central haymow with sheds on three sides. It measures 67' 2" by 64' 4" and is 31' high at the ridge.

The gambrel-roofed **new barn** was built as a hay barn in 1963 by Phil Funk, a handyman who was also the skipper of Fred Zylstra's yacht. It measures 67' wide by 100' long and is 30' 6" high at the ridge. There are mangers, or feeding troughs, extending along one side.

S18: Gordon Buchanan Farm (Red Mill Farm) (290 Valley Farms Road). Originally homesteaded by Henry Quinlan, a single man who died intestate in 1880, this farm's patent was obtained by his heirs (1882) and eventually purchased by Peter Gorman, one of several Irish immigrants who settled in San Juan Valley. The Gorman family farmed there until they sold the property to Gordon Buchanan, who ran it as a dairy farm, though he did grow other crops. In the early 1960s, after pea farming died out, Ernest and Dodie Post Gann purchased additional acreage, renamed it Red Mill Farm, and ran beef cattle. They established a conservation easement with the San Juan Preservation Trust; after Dodie's death, the title to the 800-acre farm went to the trust, which farms the land through a tenant farmer and uses the farmstead for retreats and as a native-plant and grass-seed nursery called Salish Seeds. The farmstead was sited atop a rocky area with a commanding view of the surrounding fields and countryside. Historic elements include two barns as well as a granary, milk house, and smokehouse.

The most visible structure is the **old barn**, which probably dates from the 1890s. Oriented east-west, it consists of a 30' 6"–wide-by-50'-long-by-30'-high haymow with 18'-wide sheds on the west and north sides. The west shed, which had a concrete floor added in the 1920s, contained stanchions, reflecting its use as a dairy barn. There is a walkway between the mow and the north shed, which contains horse stalls today.

Gordon Buchanan built the **new barn** in the 1930s from recycled timbers and driftwood; he only had to purchase the hardware. The timber-frame structure has bents that form four bays, with a central 26'-wide-by-52'-long-by-31'-high haymow with two 16'-wide wings. The sills are placed on stumps on top of fieldstones. Entry is by means of a central sliding door and two side doors, all in the east gable.

The **granary** is 15' wide by 36' 6" long and consists of two parts: a small room with inside-out construction (with the structure exposed in parts), with the usual "tight" construction, to prevent the intrusion of vermin, that probably had grain in bins; and a larger room for processing the grain and storing it in sacks. The **milk house** is approximately 9' wide by 10' 4" long, with shiplap siding and a 24"-tall stem wall. The 10'-tall **smokehouse**, 6' wide by 8' long, is constructed of board-and-batten siding.

S19: Peter Jewell Farm (Fir Oak Farm). Originally homesteaded by Peter Jewell, this property was later owned and farmed by John A. Doyle. Located on rocky ground overlooking fields to the north, Fir Oak Farm has a large complex of farm buildings: farmhouse, granary, tool/pump house, milk house, bunkhouse, outhouse, garage, and outbuilding. There also used to be a 30'-wide-by-60'-long barn, with a central haymow and stanchions for milking dairy cows on either side.

The **farmhouse**, which sits prominently in the middle of the farmstead, is one-and-a-half stories, with a side-facing gable roof. Like many of the older farmhouses, additions to the back incorporated a kitchen and bathroom when indoor plumbing became available. Nearest the north field is the **granary,** 18' wide by 28' long and built on wooden piers that keep it off the damp ground. Like most granaries, it is tightly built with a frame construction and shiplap sheathing; the only openings are two screened vents in the gables and large sliding doors on one side. The 16'-wide-by-33' 6"–long **tool/pump house** is located near the house; it is divided between a small room for the well and pump and a larger workshop. The **milk house**, which is almost square (10' by 12'), is also located near the house. Across the farm lane is the rare instance of a surviving **bunkhouse**, 12' wide by 16' long, with shiplap siding. A partition wall divides the structure into two rooms, with a centrally located brick chimney for a wood stove to heat both rooms. A few paces away is a "two-holer" **outhouse**, 4' 3" wide, 6' 7" long, and 8' 3" tall. Nearer to the road is a newer 18'-wide-by-20'-long **garage**. There used to be a chicken house on the farmstead, but it burned down in the 1990s. The remaining **outbuilding**, which appears to have been a poultry house of some sort, is 12' wide by 20' long with a 9' 9"–high shed roof sloping north; it is of box construction: vertical wooden boards nailed to a sill below and a plate above.

S20: Valley View Farm. In 1917 Roy Paul Guard and his wife Margaret Myrtle Hemphill bought Valley View Farm, north of the intersection of Valley Farms and San Juan Valley Roads—land that was originally homesteaded by Peter Jewell. In 1950, Clyde Madden Sundstrom and Ruth Marjorie Guard—Roy and Myrtle's daughter—bought the property and operated a dairy there as well as raising oats, barley, and hay, livestock (sheep and beef cattle), and poultry (chickens and turkeys). The farmstead has several buildings dating from the historic period of farming: a farmhouse, two barns, a milk house, and the remains of two concrete block silos.

Highest on the raised area is the **farmhouse**, which has been added on to and remodeled several times, it is *T*-shape in plan and gable roofed. Across from it is the **older barn**, nicknamed the "Barn Marché," which has a 20'-wide-by-30' 10"-long haymow with sheds on the north and east sides. The newer, **1933 barn** is a lot larger—80' long, 68' wide, and 36' tall. It has a broken-gable roof: two slopes per side, the upper one being of steeper pitch. To one side of the vast haymow are the whitewashed stanchions where the family milked the cows. This barn is listed on the Washington Heritage Barn Register.

Near this milking portion of the barn is an 8'-wide-by-10'-long **milk house**, typically constructed with a 24"-high concrete sill, wood-frame shiplap above, a shingled gable roof, two screened windows, and a door with an overhang. Two abandoned concrete **silos** stand behind the milk houses and between the two barns. They measure 10' in diameter on the inside and stand about 18' high. They are constructed of 10½"-wide-by-5"-tall curved concrete segments that interlock and are held in place by 3" metal bands. The silage was used for the dairy operations.

S21: John Sweeney Farm. Originally homesteaded in 1879 by John Archambault, a French Canadian, this property was purchased in 1890 by John Sweeney, who came from Orcas Island. Sweeney built the original **barn**, but after William Buchanan purchased the place in 1918, he raised the wooden timber-frame structure onto its current stone first floor. The haymow is accessed by means of a dirt and stone ramp (including a buried tractor!) that leads to the center drive. The gable-on-hip structure measures 75' wide by 110' long and is 35' high. In addition to the haymow on the second floor, the lower floor was used for milking cattle—complete with stanchions and a manure gutter—as well as stalls and some machine storage. It is listed on the Washington Heritage Barn Register.

Nearby are a milk house, granary, silo, and a dwelling for a hired hand. The **milk house** is approximately 8' wide by 10' long and is semi-attached to the barn with a roof overhang that covers a 3' passageway; it contains a water trough for cooling the milk in cans. A **granary**, apparently built around the same time as the milk house (1918), has a post-and-pier foundation with board-and-batten siding. The modern (1956) "Butler Building" metal **silo** rests on a concrete base; this was a common post–World War II ready-made structure. The fourth outbuilding was constructed in 1952 to house a hired hand; the **house** has two rooms and has a frame construction with wood siding and a cedar-shake roof.

S22: Tommy Davis Barn and Farmhouse. Prominently sited on a slope overlooking San Juan Valley, this **barn** was built by Thomas ("Tommy") K. Davis ca. 1927, in part with lumber from the old mill at Argyle. It was designed as a dairy barn, with a lower concrete-and-brick ground floor used for milking cows and an upper wood-frame loft used for hay storage. The plan and design were standard at the time; various barn equipment manufacturers and even lumber companies featured similar designs and even kits of lumber and other materials for constructing barns. The structure is 36' wide by 72' long, with an 18' shed on the downhill, or west, side. Under its Dutch gambrel roof is a metal hay rail-and-trolley system with a hay hood and door on the north side. At one time, on the south side of the structure there was also a pea viner with a silo to hold the vines.

Nearby but hidden by the hawthorn hedges is the hipped-roof **farmhouse**, which is almost square in plan. It is one of the few houses totally built out of cement "cast stone," an early masonry block that was made by pouring concrete into molds with a variety of faces, including rock. The building has been added onto and modified significantly.

S23: Beaverton Valley. The intersection of No. 2 Schoolhouse Road and Beaverton Valley Road affords a glimpse up Barnswallow Lane toward two farmsteads that used the rich bottomland soils and lush grass of Beaverton Valley to raise oats and provide pasture for dairy cattle. The Guard family had several farms here on the west end of Beaverton Valley.

Originally homesteaded by William Harrison Higgins in 1881, the **Paul Guard Farm** includes a house, barn, granary, and log cabin. The **log cabin**—a typical "starter house" that was later used as a root cellar—is the oldest structure on the farm. It is constructed of thick logs that have been hewn flat and connected with half-dovetail notches; a newer roof has been added on top of its original low-sloped roof. The **house**, which was built as a residence after the family moved out of the log cabin, is a side-facing, gable-roofed, one-and-a-half-story frame structure with a wing in the back to form a *T*-shaped plan. The **barn**, which is oriented east-west, consists of a 26'-wide-by-41'-long lofted space with a 17' shed on the north side that was used for milking. There is a hay rail-and-trolley system in the loft and the hay door in the west gable slides vertically—a rare occurrence in the islands. The milking shed, which is whitewashed, contains some remnants of stanchions. The gable-roofed **granary** is 16' wide by 26' long and sided with boards and battens; like most granaries, it has a single entrance in the center of one side wall and vents in the gables.

Barnswallow Farm was farmed by Harold Guard, grandson of Paul Guard, with a dairy operation that included most of Beaverton Valley. In addition to the house, which has been extensively remodeled, only one structure remains from the historic dairy operations on the farm: a **milk house**. It is a typical 8' wide by 10' long, has shiplap siding, and a 7'-shed addition to one of the gable sides. The **barn**, which was recently demolished, was oriented north-south and had a broken-gable roof that was 47' 6" wide by 50' 6" long and 33' high, with a 17' 6"–by-20' shed on the east. The central space was a floor-to-ridge haymow, with a milking parlor with manure gutter to the west side.

Traveling north on Beaverton Valley Road, one encounters a cluster of three farms: Stanbra (Nesika), on Redwing Road; Daniel B. Shull (Dancing Seeds), to the west; and Art Gilmer, to the east. These form a family neighborhood: Daniel B. and Emma Shull's main farm with two others established by their daughter Rebecca Anna Laura and son-in-law Lawrence Glenn Stanbra and their son Howard.

S24: Stanbra Farm (Nesika Farm). This farmstead includes a barn, house, and outhouse. The **barn** was built by Peter Gorman before it was owned by Lawrence Glenn Stanbra and Rebecca Anna Laura Shull, the daughter of neighbors Daniel and Emma Shull, and is currently used by Nesika Farm. Oriented north-south, it consists of a main gable area for a haymow and a milking shed addition to the east. The mow measures 24' wide by 32' long by 21' 8" high and has a hay door with steel rail-and-trolley hay system. The milking shed has at least three stanchions. It is constructed of log posts with milled beams, braces, and rafters. The **house**, which was moved across the road from the Shull Farm and sited on a slope with a basement, is one-and-a-half stories with a shed dormer and knee braces characteristic of the Craftsman style. Nearby is the **outhouse**, from the Art Gilmer farm, a "three-holer" constructed of wood board and battens with a shingled gable roof.

S25: Daniel B. and Emma Shull Farm. Originally from Indiana, Daniel Bair Shull and his wife Emmereldes "Emma" Shade came to San Juan Island in 1895 from Port Townsend. Shull served as San Juan County Commissioner. In addition to this prosperous farm, they had a "retirement" house on Argyle Avenue in Friday Harbor. The Daniel B. Shull Farm has a large complex of buildings, including a prominent barn, house, granary, the foundation of a silo, and one of the few remaining water towers on the island. Facing the road, the **house** is multifaceted, consisting of a two-story *L* with a two-story hall in the corner, wrap-around porches, and a single-story addition in the rear.

When it was built in 1908, the **barn** was considered the largest and most modern on San Juan Island, with storage for 120 tons of hay. Oriented north-south and projecting perpendicular to the slope, this is a five-bay timber-frame barn. The central haymow is 30' wide by 60' long by 32' high, with a steel hay rail-and-trolley. The side bays are 18' wide; those on the south had cow stanchions while those on the north had horse stalls. There is also a 38'-wide-by-18'-long extension to the main gable.

The **granary**, which was converted into a chicken coop, is 16' wide by 26' long, with board-and-batten siding. Near the barn, a raised concrete foundation is all that remains of a **silo**. The **water tower**, like many in the islands, is two stories tall, with the ground floor devoted to the well and pump and the top floor to the water tank itself. The floor plan of the ground floor is 12' 4" wide by 24' 6" long, with a 5' 6" shed addition. The second floor is smaller—about 12' 4" square.

S26: Art Gilmer Farm Barn. Arthur "Art" Francis Gilmer owned several farms on San Juan Island. This farm may have been run by Howard Shull, the son of Daniel B. and Emma Shull, who farmed across the road. He was featured in the 1922 article in the *Friday Harbor Journal*, "Howard Shull Operates a Model Dairy Farm." In addition to a barn, the article also mentions a separator room in the pump house and a hog barn, neither of which are extant. At 52' wide and 60' long, the simple gable-roofed **barn** is one of the largest loft barns in the islands. It has two stories: a 7' 9"-high ground floor and a 27' 8"-loft that was used for hay, complete with hay door and hood and a track-and-trolley system. The barn is oriented east-west, and the south side of the ground floor is whitewashed, with stanchions for milking dairy cattle.

S27: Bert and Jeri Lawson Farm (Amaro Farm). This farm was purchased by Alfred Lawson in 1909, and he and his wife Esther deeded it to their son Gilbert "Bert" Joseph Lawson in 1941. Bert married Adah Geraldine (Jeri) Halvorsen and they had three children: Ruthie, Richard, and Victor. The Lawson Farm was a subsistence operation, with some cash crops such as butter, cream, eggs, and beef. They raised cows, pigs, sheep, chickens, and ducks; hayed; and grew barley and wheat. They also had a large kitchen garden, and Jeri canned all their fruits and vegetables as well as meat.

The farmstead consists of a farmhouse, barn, milk house, and granary, all recently restored. The **farmhouse** is a two-story gable-roofed structure with additions. The gable-roofed **milk house**, located a regulatory 50' away from the barn, is a typical 10' wide by 12' long, with 24"-high concrete stem walls and clapboard siding. The **granary** is a 14' 4"-wide-by-22' 2"-long gable-roofed structure clad in board-and-batten siding; it has few openings—only a single door in the side and a window in each gable.

Although not completely surrounded by hips, the **barn** has sheds on three sides that form a distinctive gable-on-hip roof type. Oriented east-west and built into the hillside to take advantage of the slope by forming different floor levels, the main haymow under the gable is 30' wide by 52' long by 36' high. It has a ventilator as well as a wood rail with steel trolley, a rare example of a central hay trolley system (i.e., without a hay door and hood) in the islands. Each of the sheds is 20' wide, with the west one being whitewashed and containing 10 stanchions for milk cows. It is listed on the Washington Heritage Barn Register.

S28: Mount Grant. If you have the time, hike up the mountain that was named Mount Grant on the 1874 township and range survey; the experience is well worth it. From the 736'-summit, on clear days you will see a panorama of San Juan as well as other islands. To the south, San Juan Valley stretches out before you; the drainage to False Bay is clearly visible, and the many farms scattered through the landscape are a rich sight. Closer at hand is the Bert and Jeri Lawson Farm with its wonderful assemblage of buildings, in particular the iconic barn.

Cady Mountain, which lies to the north of Mount Grant, was named after John Keddy, who applied for a preemption claim for 160 acres just to the north, across West Valley Road, in 1877. Keddy was born a British citizen ca. 1833 near Exeter, Ontario.

The 1870 US federal census noted that he had 160 acres of land, 20 improved, 40 in woodland, and 100 unimproved, with a cash value of $800, $10 in farm implements, $360 in wages paid, one horse, and 1,000 sheep worth $2,000. The township and range survey, which surveyed and mapped the area in 1874, mentioned his house, several agricultural structures, pasture, and fences. John, together with his brother William on Lopez Island, had crop-sharing "contracts" for running sheep; under this system, a shepherd would manage 200 sheep, earning half of the wool crop and half of the increase in lambs, less any loss. By 1900 John Keddy, who was described as "English," a member of the Canadian Presbyterian Church, with the occupation "gentleman," had moved back to Ontario. He died in Brandon, Manitoba, on January 15, 1907. The mountain where John Keddy ran his sheep, however, became known as Cady Mountain.

Alfred Lawson Family Land. Most of the nearby land that you can see from Mount Grant once belonged to Alfred Lawson and his family. Born to Peter Lawson and Fannie Dearden, the daughter of the neighbor to the north, George Dearden, Alfred married Esther Lucretia Fowle on August 2, 1905. They had six children: Thomas Alfred; Marguerite Elizabeth; Howard; John Roy; Gilbert Joseph; and Theodore Glenn. Alfred Lawson started off with a 160-acre homestead to the southwest of here in 1891. In 1897 he obtained title to another homestead of 160 acres to the north. Four years later he bought an additional 20 acres, and then in 1909 he bought James Ross's homestead to the east—the Bert Lawson farm and barn—from Victor J. and Fannie Capron for $950.

S29: West Valley. As you head north down the slopes of Mount Grant and Cady Mountain, you will see to the west a large valley that doesn't have a given name but I refer to here as West Valley, like the road. Originally the Hudson's Bay Company situated a station here as part of its Belle Vue Sheep Farm. Later, several farms were established in the area. During more recent times, a trio of newer farms (Lacrover, Mitchell Bay, and Sweet Earth) have raised on this land market vegetables, chickens and eggs, nursery stock, and livestock such as sheep and pigs.

S30: Alfred Lawson Farm Barn (Sweet Water Farm Barn). Heading north, you will see a **barn** originally built by the Alfred Lawson family around the turn of the last century. It is a timber frame, with a center-entry haymow and sheds for stalls added onto the west and south sides. The barn measures 56' wide by 60' long and has a saltbox roof. Amrita and Jay Ibold bought the farm in 1999 and renamed it Sweet Water; they breed and train Akhal-Teke horses. The Sweet Water Farm Barn is listed on the Washington Heritage Barn Register.

S31: Hoffmeister Farm and Sandwith Orchard. Heading north, as you approach English Camp, to the east is an area that was farmed by Augustus Hoffmeister, who was post sutler (civilian merchant) at English Camp. Hoffmeister ran both cattle and sheep, of which some 500 were Southdowns that he had on his home ranch near the camp as well as on Spieden Island, just to the north of Roche Harbor. When Hoffmeister died in 1874 (his was the first probate filed in the newly formed San Juan County), his estate was divided between American Isaac Sandwith, who leased his property on San Juan, and Canadian John Tod Jr. (son of Hudson's Bay Company Director John Tod), who purchased the sheep, farm, livestock, and improvements on Spieden and Henry islands. In addition to grazing sheep and planting grain, Sandwith established an **orchard** of apple (Ben Davis variety), apricot, cherry, pear (Comice, Pound, Vermont, and White Doyenne), and plum trees. This area is currently part of the San Juan Island National Historical Park's English Camp and is being restored with stock grafted from the same varieties as the originals.

S32: English Camp/Crook Family Farm (3905 West Valley Road). At the San Juan Island National Historical Park's English Camp, you can see a restored version of the garden that was cultivated by the Royal Marines stationed at this site from 1860 to 1872. After their departure, the land was claimed as a homestead by William Crook, who farmed the land with his son James. They made use of the remaining British structures as farm buildings. In addition to raising grain and grazing sheep, they planted an **orchard** of trees that extended throughout the Parade Ground, and there are some extant Pound and White Doyenne pear trees that still produce fruit. In addition to caretaking the legacy of English Camp, Jim Crook was an avid craftsman, and he built a car-sized machine to card wool, which can be seen on display at the San Juan Historical Society's Museum of History and Industry in Friday Harbor.

S33: Bellevue Poultry Farm. To the west before you come to Roche Harbor is a sign for Bellevue Farm Road, the former site of the Bellevue Poultry Farm. Established by Tacoma and Roche Harbor Lime Company's John S. McMillin, the farm specialized in Crystal White Orpington chickens as well as geese and turkeys. McMillin advertised throughout the region, offering to ship breeding and laying birds as well as eggs and dressed poultry. There are no structures remaining that belonged to the poultry farm. Bellevue Farm currently grows Pinot Noir, Ortega, and Siegerrebe grapes and hard cider apples for Madrone Cellars & Ciders.

S34: White Point. In 1886, after selling their lime works at Roche Harbor to John S. McMillin and the Tacoma and Roche Harbor Lime Company, brothers Richard and Robert Scurr bought a ranch of 183 acres on nearby White Point (with a quarry and pot kiln nearby). They planted 500 apple and pear trees to form a substantial orchard and built a barnlike structure with a stone cellar, possibly used to store their fruit. After Richard died, Robert married Nettie Hill, who lived and farmed there until her death.

The apple orchards were replanted in the mid-1990s with cider apples, and **Westcott Bay Cider**, established by Richard Anderson, produced its first cider in 1999. In 2010, new partners Hawk and Suzy Pingree were added and **San Juan Distillery** (at the intersection of Anderson Lane and Armadale Road) was established to distill gin and other spirits. After driving past the orchards on White Point, stop by the distillery (open seasonally) for tastings.

S35: Rouleau Dairy (726 Rouleau Road). Started by Frank Rouleau in the 1910s, the Rouleau Dairy delivered raw milk from 35 cows. Because they did not have electricity until after World War II, they used a steam turbine for the brush and steam box used for sterilizing the bottles. In addition to the large **barn**, which was built around 1912 and measures 32'6" wide by 52'6" long by 32'6" high, there are **milk and root houses**.

S36: San Juan Vineyards (3136 Roche Harbor Road). Established in 1996 by Steve and Yvonne Swanberg, San Juan Vineyards was the second winery planted in the islands. It has Madeleine Angevine and Siegerrebe vines, whose grapes produce white wines; the vineyard's red vintages are produced with grapes from the Yakima Valley. The wine-tasting room is in the renovated 1895 Sportsman's Lake (District No. 22) Schoolhouse.

S37: Three Meadows Barn. The King **barn**, oriented on a north-south axis, is located on pastureland near a marsh. The structure, which has a large metal gable roof with sheds on the northeast, southeast, and south sides, is 98' long (plus the 16' 6"–wide shed to the northeast and the 17'-wide shed to the south) and 56' 3" wide (plus another 14' for the southeast shed). The main part of this post-and-beam, center-drive barn consists of seven bays that enclose a 33'-high haymow, serviced by a rail-and-trolley system. Three large metal ventilators are located on top of the roof. The south shed, probably added in 1934 (as suggested by an inscription in the concrete floor), shelters a milking parlor with 10 stanchions, complete with a concrete floor, manger, manure gutter, and walkways, as well as an indoor creamery in the southwest corner. The other sheds were used for stables. It is listed on the Washington Heritage Barn Register.

S38: John King Farm Barn. John William King was born in Yamhill County, Oregon, and came with his parents Francis and Sarah King to San Juan Island around 1870. He and his wife Marcia bought land here in 1900, and in 1915 they built this **barn**. Situated parallel to the slope and oriented east-west, it is approximately 60' wide by 40' long; two 18' sheds were constructed on either side of a 24'-by-40' haymow, forming a broken-gable roof.

S39: Friday Harbor Barns. Returning to Friday Harbor, you can glimpse two small barns that remain from the farms that once surrounded the town.

The **Frank Doyle Farm,** on Larson Road, is a surviving example of a small operation at the edge of the town limits. Francis Joseph Doyle, son of John P. Doyle and Mary Sweeney, was the sheriff of San Juan County in the 1930s. The **house** on the site is a "four square": two stories topped by a pyramidal roof, with inset porches in the northeast corner on both floors. The **barn** is oriented north-south and consists of two parts: a 15' 8"–high gable-roofed portion measuring 14' wide by 18' long and an 8' 6"–shed addition, accessed by double doors, forming the saltbox to the east. The shed was used for milking.

Joseph Groll was a prominent builder and lumber supplier on both Lopez and San Juan Islands and was elected to the San Juan County Commission in 1900. The Joseph Groll **barn** can be seen on the west side of the road while driving along Tucker; it is tucked behind a house and other buildings. Oriented approximately east-west, it appears to be a narrow gable-roofed barn with a central 16'-wide-by-20' long haymow and a 10' shed to the south.

Suggested Reading

Instead of footnotes, the following publications and websites are offered as suggested reading for those who wish to explore the basis of this work as well as pursue the subject further. They follow the general outline of the book: **physical features** (climate, geology, and soil); **early history of the islands** (Coast Salish, Hudson's Bay Company, explorations and surveys; and censuses); **historical archives** (land records, museum holdings, governmental archives, and newspapers); and **agricultural histories** (theses and dissertations, publications, surveys, and reports).

For information on the physical basis of agriculture in the islands, consult Earl L. Phillips, *Washington Climate for These Counties: Clallam, Jefferson, Island, San Juan, Skagit, Snohomish, Whatcom* (Pullman: Washington State University, Cooperative Extension Service, College of Agriculture, 1966). In 1975 Robert Russell edited *Geology and Water Resources of the San Juan Islands, San Juan County Washington (Washington Department of Ecology Office of Technical Services Water Supply Bulletin No. 46*), which has a lot of information on both geology and water and includes climate. Ned Brown's *Geology of the San Juan Islands* (Bellingham: Chuckanut Editions, 2014) is the latest of several attempts to describe the complex history of the islands' geology. The original *Soil Survey, San Juan County Washington* (Series 1957, No. 15) was issued in November 1962 by the US Department of Agriculture Soil Conservation Service in cooperation with the Washington Agricultural Experiment Station; the most recent (2006) survey can be accessed online at https://www.nrcs.usda.gov/Internet/FSE_manuscripts/washington/WA055/0/SanJuanWA.pdf. Two general works on the islands' climate are David Laskin's *Rains All the Time: A Connoisseur's History of Weather in the Pacific Northwest* (Seattle: Sasquatch Books, 1977) and Clifford Mass's *The Weather of the Pacific Northwest* (Seattle: University of Washington Press, 2008).

Wayne Suttles has provided the basic background of Coast Salish agriculture in the islands with his 1951 University of Washington PhD dissertation, *Economic Life of the Coast Salish of Haro and Rosario Straits*. There are several articles on Coast Salish agriculture on HistoryLink.org, the online encyclopedia of Washington State history, including: Russel Barsh and Madrona Murphy's *Coast Salish Camas Cultivation* and Russel Barsh's *Coast Salish Woolly Dogs*. For more on woolly dogs, see Candace Wellman, "Skexe: 'Birch' and Other Wool Dogs of the Coast Salish," *Journal of the Whatcom County Historical Society* (October 2001) and Russel L. Barsh, Joan Megan Jones, and Wayne Suttles, "History, Ethnology, and Archaeology of the Coast Salish Woolly-Dog," *Proceedings of the 9th Conference of the International Council of Archaeology* (Oxford: Oxbow Books, 2006).

The Hudson's Bay Company played an important role in setting the trend for EuroAmerican agriculture in the islands. For primary sources, see my transcriptions of the *Belle Vue Sheep Farm Post Journals*, Robert Firth's *Diaries*, and selections from the company's correspondence, all available online at https://www.nps.gov/sajh/learn/historyculture/belle-vue-sheep-farm-journals.htm). Richard Somerset Mackie's *Trading beyond the Mountains: The British Fur Trade on the Pacific 1793–1843* (Vancouver: University of British Columbia Press, 1997) provides an overall context.

The journals, notes, and reports of the Northwest Boundary Survey of the US Commission are invaluable to an understanding of the islands at the cusp of the Pig War. These are summarized in the *Geographical Memoir of the Islands between the Continent and Vancouver's Island in the Vicinity of the Forty Ninth Parallel of North Latitude*. In particular, see Appendix C: Report of George Gibbs, Geologist, of an Examination of San Juan Island and of the Cowitchin Archipelago and Channel; Appendix D: Report of Henry Custer, Assistant, of a Reconnaissance of San Juan Island, and the Saturna Group; and Appendix E: Report of Dr. C. B. R. Kennerly of a Reconnaissance of the Haro Archipelago. The (unedited, and therefore more opinionated) diaries and journals of Kennerly and Gibbs are also insightful.

The US government began a census of agriculture as part of the 1820 decennial census, when US marshals asked how many people within each household were engaged in agricultural pursuits. In 1840, the government began using separate census schedules to collect data related to agriculture. The census of agriculture continued to be conducted during the same year as the decennial census of population until 1950. Between 1954 and 1974, the US Census Bureau conducted the census of agriculture in years ending in the numbers four and nine. Following the census of agriculture in 1978, the Census Bureau and the US Department of Agriculture decided to conduct this census in years ending in two and seven. For the San Juan Islands, there exists agricultural data starting in 1870 and continuing each decade until 1950, with special censuses in 1925 and 1935. After 1950, there are censuses for 1959, 1964, 1969, and 1974; then in 1978, 1982, 1987, 1992, 1997, 2002, 2007, 2012, and, most recently, 2017.

The Bureau of Land Management has the records of the General Land Office, which conducted the initial township and range surveys in the islands and recorded homestead and preemption applications. Online searches will yield cadastral survey maps and basic land claim applications, but the full "proving up" records are only available through the National Archives. The Shaw Island Historical Society has the full set of these homestead papers for all the claimants on Shaw Island, and the other museums have a handful of their island papers.

The Lopez Island Historical Museum has the correspondence between Charles Edward ("C. E.") Cantine and his son, E. J. ("Ed"), from 1894 to 1904, an invaluable source on early fruit raising in the islands. The Orcas Island Historical Museum has several primary sources, including Ray Kimple's manuscript article on E. L. Von Gohren, dated October 21, 1986. The San Juan Historical Museum has several diaries and journals of early settlers, including A. J. Ackley, Henry Bailer, and Thomas Fleming. Gordon Keith compiled and edited *The Diary of James Francis Tulloch, 1875–1910* (Portland, OR: Binford & Mort, 1978)—although this might better be described as a memoir.

The Northwest Regional Branch of the Washington State Archives at Western Washington University in Bellingham contains several archives relevant to the history of agriculture in the San Juan Islands. The Territorial Court records, contained within the *Frontier Justice* collection, include many descriptions of farms and farming, while the probate files often have valuable (to the researcher) estate inventories. San Juan County records include the early *Commissioners Journals* as well as personal property assessments.

Numerous local and regional newspapers were published during the historic period covered in this book, primarily the *Islander* and the *San Juan Islander* (both available online at http://chroniclingamerica.loc.gov/lccn/sn88085190/issues). Other early Territorial regional newspapers can be found in Washington State's digital archives: http://www.digitalarchives.wa.gov/Home. Two special supplements should be noted: *The San Juan Islands Illustrated Supplement to The San Juan Islander* in 1901 and the July 19, 1908, "Islands of San Juan County, Washington" special edition of the *Everett Morning Tribune*. The *Friday Harbor Journal* (available at the San Juan Historical Museum) is also a wealth of information; in addition to historic data, there have been several articles in this newspaper in the last 50 years on the history of agriculture in the islands. Of signal interest is the series written by Nancy Larsen, "On the Road with Gordon and Clyde," several of which describe visits to farm families and their farms.

A general introduction to the early EuroAmerican agriculture in the region is Michael Leon Olsen's University of Washington PhD dissertation *The Beginnings of Agriculture in Western Oregon and Western Washington* (1970) as well as Richard White's *Land Use, Environment, and Social Change: The Shaping of Island County, Washington* (Seattle: University of Washington Press, 1980) and James R. Gibson's *Farming the Frontier: The Agricultural Opening of the Oregon Country 1786–1846* (Seattle: University of Washington Press, 1985). For an interesting comparison with British and Canadian practices across the border, see Richard Somerset Mackie's University of Victoria MA thesis, *Colonial Land, Indian Labor, and*

Company Capital: *The Economy of Vancouver Island, 1849–1858* (1984) and Ruth Wells Sandwell's Simon Fraser University PhD dissertation, *Reading the Land: Rural Discourse and the Practice of Settlement, Salt Spring Island, British Colombia, 1859–1891* (1997).

For a general history of the islands, see Lucille S. McDonald, *Making History: The People Who Shaped the San Juan Islands* (Friday Harbor, WA: Harbor Press, 1990); David Richardson, *Pig War Islands* (Eastsound, WA: Orcas Publishing Company, 1971); and Jo Bailey-Cummings and Al Cummings, *San Juan: The Powder Keg Island* (Friday Harbor, WA: Beach Combers: 1987).

During the immediate post–World War II period, there were several significant studies of agriculture in the San Juan Islands. In 1952-1953, J. C. Sherman, T. M. Griffiths, and J. L. Simms taught a geography class at the University of Washington that included field work on San Juan Island. The papers resulting from this class's study include Henry H. Davis's "Part Time Farming on San Juan Island" (August 1952); "Farm Water Supply Census," compiled by George H. Smart in the summer of 1953; and two maps—a summary map titled "Land Uses of San Juan Island" (July 1953) and a more detailed map of San Juan Valley by Duilio Peruzzi, also compiled and drawn in the summer of 1953. "San Juan County Agriculture," part of the County Agricultural Data Series, was published by the Washington State Department of Agriculture in 1956. The original 1962 *Soil Survey, San Juan County Washington* mentioned earlier contains a summary of farming in the islands. Elizabeth E. Ellis's 1978 paper, "The History of San Juan County Agriculture with an Appendix of Excerpts from the Annual Narrative Reports Filed by the County Agents" (written as part of Agriculture Economics 301 in spring 1978, taught by Dr. Harrington) is invaluable not only for the information she provides but also the precis of the reports of the county agents—some of which may have been lost subsequent to her research.

Based on a countywide survey begun in 2009, a description of the various extant historic barns in the San Juan Islands, as well as their history, design, and preservation, can be found at the 100 Friends of Old Island Barns' website "Historic Barns of the San Juan Islands": https://historicbarnssanjuanislands.com/.

Several recent studies shed light on the current state of agriculture in the San Juan Islands. In 2011, the San Juan County Agricultural Resources Committee, San Juan County Land Bank, San Juan Preservation Trust, and San Juan Islands Conservation District prepared *Growing Our Future: An Agricultural Strategic Plan for San Juan County, WA*. As a follow-up to the recommendations in that report, the San Juan County Agricultural Resources Committee obtained a USDA Local Food Promotion Program grant to conduct the *San Juan County Food Hub Feasibility Study: Coordinating Sales and Distribution of Food Grown in the San Juan Islands, Washington* (December 2016). In order to comply with the Growth Management Act's requirements in regard to critical areas—particularly the intersection of farmland with wetlands—the San Juan Islands Conservation District prepared the "Voluntary Stewardship Program" (October 30, 2017). As part of the update of the comprehensive plan, San Juan County hired Community Attributes Incorporated to prepare the "Economic Analysis of Resource Lands" (December 2017), which includes Agriculture Resource Lands.

Photograph and Illustration Credits
(by institution in order of appearance)

Beinecke Rare Book & Manuscript Library, Yale University
Belle Vue Sheep Farm, photograph taken by Northwest Boundary Survey ca. 1859; Northwest Boundary Survey Photo (10 & 11)

Lopez Island Historical Museum
GEM Farm (57)
Walter Burt with Horse-Drawn Binder, Lopez Island (70)
Berry Pickers near Flat Point on Lopez Island, ca. 1920 (83)
Threshing Crew, Lopez Island (196)

Orcas Island Historical Museum
Apple and Pear Exhibit, Orcas Island (60)
Clark Farm Prune Dryer on Orcas Island (77)
Spraying Mrs. Waldrip's Orchard on Orcas Island (81)
Beck Children Feeding Turkeys, Orcas Island, 1920 (101)
Mrs. Waldrip's Water Tower, Orcas Island (115)

Roche Harbor Resort
Bellevue Poultry Farm, Roche Harbor, San Juan Island (98)

Royal Ontario Museum
"Clal-lum Women Weaving a Blanket" by Paul Kane, Showing Woolly Dog [B&W version] (9)

San Juan County Assessor's Office
San Juan County Agricultural Resource Lands and Current Use Farm and Ag (122)

San Juan County Land Bank
Orcas Island Farmer Vern Coffelt, 2007 (125)

San Juan Historical Museum
Threshing at the King Farm (cover)
Katz Farm Old Town Lagoon [cropped] (title page)
San Juan Valley from Little Mountain, San Juan Island (4)
View of King Farm with Friday Harbor in the background (55)
Draft Horse Team Hauling Stone Boat (61)
Driving Sheep down Spring Street, Friday Harbor (69)
Sheep on the Friday Harbor Dock Awaiting Shipment (65)
Lizzie Lawson Haying (67)
Harvesting Oats, Paul Guard Farm, Beaverton Valley, San Juan
    Island (69)
Orchards in Eastsound, Orcas Island ca. 1900 (73)
G. B. Driggs's Prune Orchard, corner of Malcolm and Argyle
    in Friday Harbor (78 & 79)
Pacific Northwest Guernsey Breeders Association 1922 Tour of the
    San Juan Islands (89)
Harvesting Rhubarb at Hershberger Farm, Deer Harbor,
    Orcas Island (109)
Early Farmers Market at the American Legion in Friday Harbor (126)

San Juan Island National Historical Park
Lt. James W. Forsyth Tracing of Map of South-East End of
    San Juan Island (14)
James Madison Alden Sketch of Belle Vue Sheep Farm 1859 (15)
Charles Griffin Entry in *Belle Vue Sheep Farm Post Journal* June 15,
    1859 (23)
Harvesting the Prairie at Firth Farm (American Camp) (71)

Sundstrom Family
Ruth Guard Milking (95)
John Sundstrom Working Pea Viner, Gorman Farm, San Juan Valley (103)

Washington State Library/ Washington State Digital Archives
Salish Woman Digging Roots (7)
Fences at the Firth Farm (American Camp) (53)

Washington State University Map Collections
1910 Soil Map of San Juan County [cropped and B&W] (2)

Maps by Lovel and Boyd C. Pratt;
    assistance by W. Bruce Conway
Diagrams, Drawings, and Graphs by Boyd C. Pratt

Acknowledgements

- Lovel Pratt: fellow former farmer, helpmate, true friend and the love of my life
- Readers: Elaine Kendall, Nancy McCoy, Edrie Vinson, and Ron Zee
- San Juan County Agricultural Resources Committee: Peggy Bill
- San Juan County Assessor's Office: John Kulseth and Chris Ledgerwood
- San Juan County Land Bank, for whom I researched the history of many farms: Lincoln Bormann, Judy Cumming, Ruthie Dougherty, Eliza Habegger, and Doug McCutchen
- WSU San Juan County Cooperative Extension Agents: Tom Schultz and Brook Brouwer
- Lopez Island Historical Museum: Amy Hildebrand and Mark Thompson-Klein
- Orcas Island Historical Museum: Brittney Maruska, Edrie Vinson, and Jen Volmer
- San Juan Historical Museum: Kevin Loftus and Andy Zall
- Shaw Island Historical Museum: Cherie Christiansen
- San Juan Island Library, research enabler: particularly Heidi Lewis, interlibrary loan extraordinaire
- Shaun Hubbard, Nancy Larsen, Nancy McCoy, Sandy Strehlou, Mike Vouri, and the eagle-eyed Louise Dustrude
- W. Bruce Conway, WBC Book Design services
- Jill Twist, excellent editor—but all mistakes remain my own, despite her best efforts!

Glossary

Barn. A large building used for storing agricultural crops and housing livestock.

Creamery. A place where butter and other dairy products are produced from cream.

Frame construction (also called **balloon** or **platform frame construction**). With the advent of milled lumber, support and wall systems were constructed of dimensional lumber such as 2x4s, 2x6s, 2x8s, 2x10s, etc. Because this lumber was so much smaller than timber, it was often referred to as "stick" construction. In forming a two-story structure, there were two methods: balloon, where upright studs extended the full height of the structure and the floor was "hung" from them, or platform, in which each story had a separate structure (one placed on top of the other).

Fruit dryer (also called **fruit evaporator**). A large structure, usually consisting of a lower story for heat and an upper story with screens, used to dry fruit. Most had a pyramidal roof with a vent at the apex.

Granary. A storehouse for threshed grain.

Haymow (also called **hayloft**). In early barn designs, the central section of the barn, called the mow (pronounced like "cow"), was used for storing loose hay, piled from the (dirt) floor to the peak. Later barns incorporated lofts, or second stories solely for storing hay—at first loose and later baled.

Hay rail-and-trolley system. A means of conveying loose hay into a barn. A rectangular wooden beam or metal track hung from the underside of the roof ridge, upon which a trolley could be wheeled to position over select areas of the haymow. Pulleys on the trolley were threaded with ropes that were attached on one end to either hay hooks or slings, and on the other to a draft team (and later a tractor or truck) to pull it and thus elevate the load. The loose hay was hoisted up by the pulleys on the trolley, often located under a sheltering hay hood, to the hay door, and then wheeled into the barn. The load, suspended high above the mow or loft, was released by pulling a trip rope that would disengage the hooks or sling.

Hudson's Bay Company frame. The standard form of construction used by the Hudson's Bay Company in general and at Belle Vue Sheep Farm on San Juan Island in particular. The general French Canadian term for log construction was *pièce sur pièce* (simplified from *pièces de bois sur pièces de bois*—literally translated as "pieces of wood on pieces of wood"). More specifically, structures that consisted of vertically grooved posts filled with planks or squared logs were called *poteaux et pièce collisante* (posts and sliding piece), and the posts themselves were placed on sills (*poteaux sur sole*), differing from another method in which the posts were set in the ground (*poteaux en terre*). With its dissemination into the Red River Valley by French Canadian *voyageurs*, *pièce sur pièce poteaux et pièce collisante* took on the name of Red River style. After the absorption of the Red River–based Northwest Company by the Hudson's Bay Company, the style soon became known as Hudson's Bay Company frame, where it was used throughout the West—so much so that it is also commonly referred to as the Canadian style.

Kanaka. General term for a Hawaiian islander, particularly those employed by the Hudson's Bay Company as shepherds and swineherds.

Milk house. A structure housing machinery for separating cream from skim milk and storing the cream until delivery was arranged.

Northwest Boundary Survey. A survey of the boundary (mostly along the 49th parallel) between Great Britain (Canada) and the United States that had been established by the Treaty of 1846.

Pole construction. Barn builders erected simple poles with beams attached with nails or spikes. At times these poles were supported on foundation stones or were buried into the ground. A common distinctive feature of these poles is that they are not only barked (the bark removed with a draw knife) but also charred. It is not clear if trees already burned in forest fires were deliberately chosen for barns, or if logs were charred after felling. Dimensional milled lumber, usually 2x8s, were nailed to the tops of the posts as plates, or to the sides as braces. The roof was then constructed with standard dimensional lumber.

Public Land System (also called **township and range**). A surveying system, created by the Land Ordinance of 1785 and first used in the old "Northwest" (Ohio), to plat and divide real property for sale and settling. From principal lines running north-south and east-west at six-mile intervals, surveyors platted east-west township lines and north-south range lines; townships were formed by the intersections of these lines. These townships were in turn subdivided into 36 sections (or squares consisting of one mile on each side) of 640 acres each, which could then be subdivided into half sections of 320 acres, quarter sections of 160 acres, eighth sections of 80 acres, and so on. A township and range survey of the islands began in 1874 and continued for several years, with corrections and amendments occurring in the early twentieth century.

Roof shapes. The shape of the roof is a defining feature of a building, particularly barns because different shapes allow for varying capacity for hay storage. The simplest roof shape is a **shed**: a single plane that is high on one side and low on the other. Most shed roofs do not stand alone but cover additions to a building. A **gable** roof consists of two equally pitched slopes meeting at a ridge. Farmers often added sheds (also called lean-tos) to one or several sides of their barns, giving the building a spread-out look. Depending on how the shed was added, different roof forms result. When a shed is added to one side, continuing the roof line, it might be called a **saltbox** (integrated lean-to). If added to one or two sides but with a shallower pitch, the result is called a **broken gable**. If sheds were added to both sides but started lower on the wall, thus leaving the upper part of the side exposed, this was known as a **monitor** or **western** roof. At times the roof extended all the way down to a first-floor side wall, in which case the form is historically called a **Dutch** roof. With the advent of milled lumber and standardized plans, frame walls supported roof trusses, which were in turn built up with sections of small, dimensional lumber to form the distinctive gambrel and gothic roofs that we associate with classic dairy barns. Proponents of these roof systems argued that not only did they allow for greater volume of space for hay storage, but there were no timber-frame cross beams or bents to get in the way of hay-track systems, and they used less wood, skilled labor, and handwork. **Gambrel** roofs are defined by two slopes per side; an upper one with a pitch of about 30 degrees and a lower one with a pitch of 60 degrees. Gambrel-roofed barns are called English if the eaves continue in a straight line and Dutch if they flare upward (forming a "Dutch knuckle"). The lamination of smaller pieces of lumber to form continuously curved arches led to the **Gothic**, or pointed, roof.

Root cellar (also called **root house**). A structure, usually partially underground, used to store root vegetables such as carrots, potatoes, and turnips.

Silo. A towerlike structure used to store either grain or silage (grass or other green fodder compacted and stored in airtight conditions, without first being dried, and used as animal feed in the winter).

Smokehouse. A structure used for preserving food (usually meat or fish) by exposure to smoke.

Stanchion. Usually located in sheds abutting the barn's haymow, with stalls, and were used to milk dairy cattle twice a day. A typical stall section included an alley or walkway for carrying feed to the mangers, the manger itself, a stanchion for holding the milch cow in place while it was being milked, a platform for the animal to stand on, and then another alley or walkway with a gutter or trough for the manure that the cow dropped while being milked.

Timber-frame construction. Hewn posts and beams connected with pegged, mortise-and-tenon joints. Large sills, or horizontal beams resting on stone or stone-and-wood-pier foundations, supported upright posts and horizontal beams called plates (on top of the posts) or girts, spanning between. Typical timber sizes were around 8" x 8", although they could be either smaller or larger.

Smaller (6" x 6") timbers, called braces, were placed diagonally in the corners of the frame to prevent wracking. The roof was supported by a cross tie (also called a cross beam or cross girt), which formed the lower tie of a truss that was supported by either a single ("king") post or two ("queen") posts. The rafters above formed the two top sides of the triangle. Horizontal beams running the length of the roof were called purlins, and diagonal braces supporting these from the crossbeams were called struts.

US Coast and Geodetic Survey. First established as the US Survey of the Coast (1807) and then the US Coast Survey (1836–1878), the purpose of the Coast and Geodetic Survey (1878–1970) was to map the coastlines and chart the waters of the United States. Teams surveyed the San Juan Islands from 1888 to 1897.

Ventilators. Ventilators are included in barns for two reasons: to allow for warm air to escape, in order to keep a lower space cool (used in fruit barns and root cellars); and to allow for the ventilation of humid and odiferous vapors, usually from dairy or poultry areas. The location of the ventilation shaft could either be in the approximate middle of the floor plan, up through the ceiling and roof structure, or, in some cases, following the side of the barn upward along the underside of the gable or gambrel roof to a ventilator located at the peak. Early ventilators were constructed of wood framed around a one-foot-square shaft, with a gable roof and wooden louvers on the sides to allow for ventilation while discouraging the entry of vermin. In the case of larger and more elaborately decorated ventilators, one could easily call them cupolas. In the early twentieth century, with the introduction of prefabricated dairy equipment, galvanized-steel ventilators became available; these ventilators featured fans on the inside that rotated with the drafts of rising warm air.

Water tower. A tower supporting an elevated water tank, whose height creates the pressure required to distribute the water through a piped system.

Plants and Animals

Native Trees
Bigleaf maple (*Acer macrophyllum*)
Douglas fir (*Pseudotsuga menziesii*)
Garry or white oak (*Quercus garryana*)
Lodgepole pine (*Pinus contorta*)
Madrona (*Arbutus menziesii*)
Red alder (*Alnus rubra*)
Rocky Mountain juniper (*Juniperus scopulorum*)
Sitka spruce (*Picea sitchensis*)
Western hemlock (*Tsuga heterophylla*)
Western red cedar (*Thuja plicata*)
Western white pine (*Pinus monticola*)
White fir (*Abies grandis*)

Native Grasses
California oatgrass (*Danthonia californica*)
Idaho fescue (*Festuca idahoensis*)
Junegrass (*Koeleria cristata*)

Plants Cultivated by Coast Salish
Death or poison camas (*Zigadenus venenosus*)
Great camas (*Camassia leichtlinii*)
Small or purple camas (*Camassia quamash*)
Chocolate lily or rice root (*Fritillaria affinis*)
Columbia or tiger lily (*Lilium columbianum*)
Harvest or crown broadiaea (*Broadiaea coronaria*)
Wild or Indian carrot (*Conioselinum pacificum*)

Flora Used by Coast Salish
Fireweed (*Chamerion angustifolium*)
Ironwood (*Holodiscus discolor*)
Nettle (*Urtica dioica*)

Animals Used for Farming by Coast Salish
Mountain goat (*Oreamnos americanus*)

Flora Introduced by EuroAmericans as Cultivars
Annual or Italian ryegrass (*Lolium multiflorum*)
Bentgrass (*Agrostis* spp.)
Clover (*Trifolium* spp.)
Downy chess or cheatgrass (*Bromus tectorum*)
Early hairgrass (*Aira praecox*)
Kentucky bluegrass (*Poa pratensis*)
Meadow foxtail (*Alopecurus pratensis*)
Orchardgrass (*Dactylis glomerata*)
Perennial ryegrass (*Lolium perenne*)
Red fescue (*Festuca rubra*)
Redtop (*Agrostis alba*)
Reed canary grass (*Phalaris arundinacea*)
Ripgut (*Bromus rigidus*)
Silver hairgrass (*Aira caryophyllea*)
Soft chess (*Bromus mollis*)
Tall fescue (*Festuca arundinacea*)
Timothy grass (*Phleum pratense*)
Velvet grass (*Holcus lanatus*)
Vetch (*Vicia* spp.)

Native Animals Considered by EuroAmericans to Be Pests
Bear (*Ursus* spp.)
Deer (*Odocoileus virginianus*)
Mink (*Mustela vison*)
Wolf (*Canis lupus*)

Animal Introduced by EuroAmericans Considered to Be a Pest
Norway or brown rat (*Mus decumanus*)

Native Plants Considered by EuroAmericans to Be Weeds
Black hawthorn (*Crataegus douglasii*)
Nettle (*Urtica dioica*)

Plants Introduced by EuroAmericans Considered to Be Weeds
American black nightshade (*Solanum nodiflorum*)
Annual sowthistle (*Sonchus oleraceus*)
Buckhorn plantain (*Plantago major*)
Canadian thistle (*Cirsium arvense*)
Common hawthorn (*Crataegus monogyna*)
Dock or sorrel (*Rumex acetosella* or *R. crispus*)
European foxtail (*Alopecurus carolinianus* and *Hordeum*)
Knotweed (*Polygonum nigrum*)
Lambsquarters or hogweed (*Chenopodium album*)
Low speargrass (*Poa annua*)
Mayweed (*Anthemis cotula*)
Pigweed (*Amaranthus spp.*)
Prostate knotweed (*Polygonum aviculare*)
Scotch broom (*Cytisus scoparius*)
Shepherd's purse (*Capsella bursa-pastoris*)

San Juan County Extension Agents

Russ Turner, one week per month, 1919

William Arthur Ness, first full-time San Juan County Extension Agent, 1921

William W. Henry, 1921–1923

H. J. Lechner, 1923–1926

Ben Beach, 1927–1933

Charles T. Meenach, 1934–1937

C. F. Webster, 1937–1940

Floyd F. Svinth, 1940–1943

W. Jim Wylie, 1943–1947

William W. Baker, 1947–1962

John G. Westergreen, 1962–1965

Edgar F. McMinn, 1964–1967

Joe Long, 1967–1969

Burrell D. Osburn, 1969–1980

Lee Campbell, 1980–1993

Tom Schultz, 1993–2016

Brook Brouwer, 2016–

Island Farmers

These biographical sketches of the men and women who farmed in the San Juan Islands are not meant to be comprehensive; only farmers mentioned in either the history or the tour guides, or both, are included. There are farmers, past and present, who are not named here. My apologies to them and their families for not including everyone who played a significant role in the history of farming in the islands.

Birth and death dates, where known, are given, as is marriage data. Farming has always been preponderantly a family business, involving husband and wife and their children—all hands being needed. Historically, however, family farms have, as a rule, been identified with men—as the "head of the household"—even though women were often partners in the farming operations. Homestead dates are the year the patent was issued (i.e., title), not the date settled or claimed, and in some cases after the death of the claimant—a source of confusion.

A. F. (Alexander Frye) Ackley (1843–1929): from Maine; married *Annie Mulno* (1853–1922) in 1871; farmer, Mulno Cove, San Juan Island.

Mike (Michael) L. Adams (1830–1913): born in Pennsylvania; nurseryman, Eastsound, Orcas Island.

George H. (Hubert) Adkins (1848–1933): born in Wisconsin; married *Mary E. Bowen* (1856–1940) in 1884; orchardist, Orcas Island.

John Anderson (ca. 1837–1882): a boot and shoemaker formerly of Port Ludlow, Washington; shot to death by his neighbor John Kay over a "breachy cow"; farmer, Lopez Island.

J. C. (James Ciprien) Archambault (1826–1901): French Canadian; married *Mary Delaunais* (1850–1924) in 1863, French Canadian/Cowlitz; homesteader (1883) and farmer, San Juan Valley, San Juan Island.

John Archambault (1864–1933): French Canadian; homesteader (1879) and farmer, San Juan Valley, San Juan Island.

Charles E. (Emil) Arnt (1906–1990): stockman of Charolais, Driftwood Ranch, Orcas Island.

Robert Auld (1845–1916): farmer and dairyman, Crow Valley, Orcas Island.

E. P. (Engelbert Phillip) Bailer (1826–1913): German; married **Dorothy A. Schwartz** (1839–1931) in 1871; homesteader (1884), San Juan Valley, San Juan Island.

Gregory Bailer (1817–1882): German; homesteader (1883), San Juan Valley, San Juan Island.

John H. (Hurd) Bartlett (ca. 1835–1922): farmer and owner of first steam thresher in the islands, Lopez Island.

Patrick Beigin (1837–1908): Irish; US Army soldier stationed at American Camp; naturalized 1873; married **Lucy Morris** (1848–1923), Howcan (Sitka) Tribe, in 1860; homesteader (1888), San Juan Valley, San Juan Island.

John Biendl (1885–1941): German, naturalized in 1890; arrived on Shaw (1902); married **Ruth Ellen Shaw** (1893–1959); farmer, Shaw Island.

August Bjork (dates unknown): homesteader (1891), Shaw Island.

James Blake (1826–1907): Irish; married **Elizabeth Ann Shannon** (1829–1868) in 1851 and **Hannah Lee** (1840–1879) in 1870; came to Lopez Island in 1883; farmer, Lopez Island.

T. J. Blake (1915–1998): grandson of James Blake; married **Wyvonna "Bonnie" Leunatta Flatter** (1917–2015); farmer, Lopez Island.

Thomas George Blake (1862–1942): born in Ontario; son of James Blake; married **Barbara Krumholtz** (1862–1892) in 1881 and **Olive Theresa Murphy** (1872–1957) in 1903; father of T. J. Blake; farmer, Lopez Island.

P. (Peter) Bostian (1838–1912): married ***Margaret Elwing Bender*** (1841–1933) in 1864; nurseryman and San Juan County Commissioner, Eastsound, Orcas Island.

George Washington Boulton (1837–1907): from Virginia; married ***Fannie Ann Dove*** (1843–1907) in 1862; farmer, Center Valley, Lopez Island.

Stephen Boyce (1827–1909): married ***Lucinda Elizabeth Steward*** (1836–1913) in 1856; farmer and San Juan County sheriff, San Juan Island.

Albert M. Breedlove (1849–1928): married ***Columbia Ellen Scisco*** (1862–1941) in 1879; Master of Friday Harbor Grange #225; farmer, San Juan Island.

John E. (Eberhard) Bruns (1873–1962): German; from Iowa; married ***Lillie Augusta Marold*** (1876–1963) in 1901; settled on Shaw Island ca. 1901; fruit and ginseng grower and dairyman, Blind Bay, Shaw Island.

James Alexander "Andrew" Buchanan (1851–1935): Canadian; arrived on Lopez Island in 1891; farmer and owner of a thresher, Center Valley, Lopez Island.

(Wilfred) Gordon Buchanan (1904–1995): farmer, San Juan Valley, San Juan Island.

Catherine Bull (ca. 1852–1932): daughter of John Bull, Kanaka, and Fu–hue–wut Mary Skqulap, Clallam/Lummi; homesteader (1883) and farmer, San Juan Valley, San Juan Island.

John Bull (1823–ca. 1860): Kanaka; married Fu–hue–wut Mary Skqulap (1830–1892), Clallam/Lummi; Hudson's Bay Company shepherd, San Juan Island.

John H. (Hutchison) Burt (1858–1933): born in Pennsylvania to Scots parents; married ***Sarah Jane "Jennie" Baker*** (1863–1904) in 1886; moved to Lopez Island in 1906; farmer and builder, Center Valley, Lopez Island.

Joseph Burt (1852-1941): born in Scotland; married ***Adella (Della)*** (1865-1944) in 1889; moved from Iowa to Lopez Island in 1902; farmer and finish carpenter, Center Valley, Lopez Island.

C. B. (Charles Bertram) Buxton (1869-1956): born in Virginia; married ***Mary Louise Mackey*** (1879-1964) in 1900; orchardist and San Juan County Assessor, Crow Valley, Orcas Island.

Robert M. Caines (1850-1927): born in Louisiana; married ***Margaret Douglas*** (1855-1937) in 1876; farmer and dairyman, Orcas and San Juan Islands.

C. E. (Charles Edward) Cantine (1838-1912): born in Michigan; married ***Helen Scott Wallace*** (1839-1873) in 1862; father of Edward I. Cantine; married ***Matilda Wilson*** (1838-?) in 1880; moved to Lopez Island in 1893; orchardist and San Juan County Surveyor, Lopez Island.

Edward "Ed" I. Cantine (1864-1950): born in Michigan; owned land, upon which his father C. E. raised an orchard, near Lopez Village.

Victor Capron (1867-1934): born in Rome, New York; moved to Roche Harbor, San Juan Island in 1898; medical doctor, farmer of sheep and Holstein cattle, temporary chairman of San Juan County Dairyman's Association, San Juan Island.

Albert "Bert" Earnest Murray Chalmers (1873-1932): Welsh; came with his father Alexander and older brother Estyn Murray when they immigrated to Orcas Island in 1891; married ***Dorothy Newill Smith*** (1885-1949); sold his farm in Crow Valley, Orcas Island, in the late 1910s.

Estyn M. (Murray) Chalmers (1870-1935): Welsh; came with his father Alexander and younger brother Bert when they immigrated to Orcas Island in 1891; married ***Gertrude Mercy Geoghegan*** (1875-1955); dairyman, Crow Valley, Orcas Island.

Brent Charnley (1957-) and ***Maggie Nilan*** (1962-): vintners, Lopez Island Vineyards, Lopez Island.

Hans Lee Christensen (*took his middle name, Lee, as his last name; known as both Hans Christensen and Hans Lee*) (1841–1926): Norwegian; emigrated in 1868; bought property on Shaw Island in 1879; married **Henrietta Kathrina (Catharina) Krafft Carr** (1851–1938) in 1902; farmer, Shaw Island.

W. (William) O. Clark (1857–1912): born in England; married **Ursula E. Rose** (1871–1929); orchardist, Westsound, Orcas Island.

(Theodore) Vern Coffelt (1930-2013): married **Sidney Reynolds** in 1977; farmers, Coffelt Farm, Crow Valley, Orcas Island.

James Cousins (1834–1921): brother of John; homesteader (1883), Lopez Island.

John Cousins (1851–1930): brother of James; married **Ellen Burt** (1854–1936); moved to Lopez Island in 1882; farmer, Lopez Island.

James "Jim" Crook (1873–1967): son of William; farmer, English Camp, San Juan Island.

William Crook (1837–1901): father of James; farmer, English Camp, San Juan Island.

George N. (Nelson) Culver (1835–1918): born in Vermont; married **Diana Louisa Aikens** (1836–1920) in 1861; farmer, Alderbrook Farm (fruit, Oxford Down sheep), Doe Bay, Orcas Island.

Lyman Cutlar (ca. 1834–1874): moved from Kentucky to San Juan via the Fraser River Gold Rush; farmed with his Indian wife (name and tribe unknown); the "man who shot the pig," San Juan Island.

Thomas "Tommy" K. Davis (1879–1964): born in Wales; married **Alice Lucinda Boyce** (1878–1945); dairy farmer, San Juan Valley, San Juan Island.

Alfred Douglas (1866–1947): arrived in the United States in 1870; son of Robert Douglas; farmer, San Juan Valley, San Juan Island.

James Douglas (1803–1877): born in Demerara (later Guyana); Chief Trader and then Chief Factor (and head of farming operations), Hudson's Bay Company, Fort Vancouver; founder (1849) of Fort Victoria; Governor, Colony of Vancouver Island; Governor, British Columbia; established Belle Vue Sheep Farm (1853), San Juan Island.

Robert Douglas (1827–1881): from Ontario; married **Matilda Kyle** (1833–1901) in 1851; homesteader, San Juan Valley, San Juan Island.

Francis "Frank" Joseph Doyle (1877–1954): son of John P. Doyle and Mary Sweeney; married **Katherine McCarthy** (1888–?) in 1917; San Juan County sheriff and farmer, Friday Harbor, San Juan Island.

John A. (Anglin) Doyle (1869–1960): born in New Brunswick; son of John P. Doyle; naturalized 1891; married **Bridget Emma Boyle** (1879–1967?) in 1917; farmer, San Juan Valley, San Juan Island.

John P. (Patrick) Doyle (1825–1902): Irish; born in New Brunswick; father of John A. Doyle; married **Mary Sweeney** (1834–1928), Irish, in 1855; farmer, San Juan Valley, San Juan Island.

G. B. (Granville Baber) Driggs (1857–1930): born Oregon; married **Fanny Lake** (1861–1925); fruit raiser and owner of the Blue Front Store, Friday Harbor, San Juan Island.

Joseph Ender (1861–1959): farmer, Center Valley, Lopez Island.

Andre Entermann (1980-) and **Elizabeth Metcalf** (1977-); farmers, Sunnyfield Farm (est. 2014), Center Valley, Lopez Island.

William "Henry" Allison Erb (1901–1994): married **Lavinia Jane McCauley** (1911–2005) in 1931; farmer, Lopez Island.

Robert Firth (1831–1901): born Pomona, Scotland; married **Jessie Grant** (1830–1889) in 1857; managed Belle Vue Sheep Farm until 1873; naturalized in 1878; homesteader (1884) and farmer, San Juan Island.

Thomas Fleming (1822–1907): Scottish; married **Mary Jane Matier** (1821–1902) in 1847; homesteader (1877), San Juan Valley, San Juan Island.

Elihu B. (Burrit) Fowler (1887–1943): born on Orcas Island; son of Oscar and Theresa Louise Fowler; married *Mary Elizabeth "Bessie" McGee* (1890–1950) in 1916; poultryman, Shaw Island.

Oscar William Fowler (1831–1915): born in Ohio; married *Theresa Louise Howard* (1848–1916) in 1870; father of Elihu Fowler; homesteader (1889) and poultryman, Shaw Island.

Robert Frazer (ca. 1827–1912): brother of William Frazer; homesteader (1883) and farmer, San Juan Island.

William Frazer (1818–1878): brother of Robert Frazer; homesteader (1883) and farmer, San Juan Island.

Peter Frechette (1826–1897): French Canadian/Cowlitz; homesteader (1894) and farmer, Crow Valley, Orcas Island.

Joe Friday (1851–1895): Kanaka/Cowlitz; son of Peter Friday; homesteader (1883) and farmer, San Juan Valley, San Juan Island.

Pierre (Peter) Friday (1830–1894): Kanaka; father of Joe Friday; shepherd, Belle Vue Sheep Farm, San Juan Island.

Joseph Fargie Gallanger (1866–1930): from Michigan; arrived on Lopez Island in the 1890s and married 16-year-old *Susie May Cochran* (1880–1970) of Port Stanley, Lopez Island in 1896; farmer, Lopez Island.

William Walter Gallanger (1868–1965): from Michigan; arrived on Lopez Island in 1891; married *Annie Eliza Bartlett* (1868–1946) in 1894; farmer, Lopez Island.

Ernest Gann (1910–1991) and *Dodie Post* (1922–2012): Red Mill Farm, San Juan Valley, San Juan Island.

George William Gibbs (1830–1919): English; arrived on Orcas Island in 1886 or 1887; farmer and orchardist (fruit, nuts, tulips), Orcas, Orcas Island.

Arthur "Art" Francis Gilmer (1890–1990): farmer, San Juan Island.

Byron Norbert "Barney" Goodrow (1915–2004): farmer; married *Marguerite McCauley* (1915–2010); berry picker, Lopez Island.

Joseph J. Gorman (1862–1913): son of Patrick and Ellen; born in California; moved to San Juan Island in 1869; married ***Mary Creeden*** (1859–1956) in 1898; farmer, San Juan Valley, San Juan Island.

Patrick Gorman (1827–1892): born in Ireland; married ***Ellen*** (1826–1899); moved to San Juan Island in 1869; father of Joseph, Mary, and Peter; farmer, San Juan Valley, San Juan Island.

Peter Gorman (1859–1949): Irish; son of Patrick and Ellen; born in California; moved to San Juan Island in 1869; farmer, San Juan Valley, San Juan Island.

Thomas Graham (1858–1944): Irish; brother of William; married ***Harriet "Hattie" Mary Glasscock***; farmer, Richardson, Lopez Island.

William Graham (1842–1928): Irish; brother of Thomas; married ***Mary Wilson*** (1844–1924) in 1871; farmer, merchant, and postmaster, Richardson, Lopez Island.

Charles John Griffin (1827–1874): English; born in Lower Canada; Clerk, later Chief Trader, Hudson's Bay Company Belle Vue Sheep Farm, San Juan Island.

George Oscar Griswold (1841–1913): from Illinois; married ***Mary Ann Kennedy*** in 1859; farmer, orchardist, and postmaster, Shaw Island.

Joseph "Joe" Groll (1868–1925): married ***Alice Kromer*** (1873–1924) in 1897; builder and lumber supplier (Lopez and San Juan Islands); San Juan County Commissioner (1900); farmer, Friday Harbor, San Juan Island.

Frank Guard (1861–1941): born in England; married ***Mary Alice First*** (1858–1935) in 1886; farmer, Beaverton Valley, San Juan Island.

Harold (James) Guard (1890–1948): married ***Emily Jane Wright*** (1893–?) in 1914; dairyman, Beaverton Valley, San Juan Island.

Leroy "Roy" Paul Guard (1886–1950): married *Margaret Myrtle Hemphill* (1889–1940); farmer, Valley View Farm, San Juan Valley, San Juan Island.

Paul Guard (1839–1915): born in England; married *Elizabeth Ann Melhuish* (1839–1911) in 1859; farmer, Beaverton Valley, San Juan Island.

William B. Hambly (1844–1936): English; emigrated in 1851; married *Mary C.* (1862–?); homesteader (1882) and orchardist, Crow Valley, Orcas Island.

Harriet Delila "Lila" Hannah (1865–1954): daughter of James Hannah; married *L. Robert Firth* (1857–1927); son of Robert Firth; San Juan Island.

James "Jim" Madison Hannah (1832–1880): from Missouri; married *Minerva Elizabeth Cahoon* (1845–1929); homesteader (1882) and farmer, San Juan Island.

Benjamin "Ben" E. (Edward) Tackaberry Harrison (1866–1945): from Wisconsin; Secretary, Orcas Island Fruit Growers Association, Eastsound, Orcas Island.

William Hayton (1878-?): married *Nellie J. Vike* (1895-1982); farmer, Lopez Island.

George Heidenrich (1861–?): poultryman (turkeys), San Juan Island.

John M. "Pea" Henry (1899–1968): from Spokane; pea farmer (San Juan Valley) and cannery owner (Friday Harbor); Mount Vernon and San Juan Island.

George L. Hershberger (Herchberger) (1859-1933): born in Maryland; married *Sylvia Haynes* (1874-1956) in 1907; rhubarb farmer, Deer Harbor, Orcas Island.

Owen James Higgins (1895–1973): married *Eva May Weir* (1903–1982) in 1923; carpenter, shipwright, and farmer, Richardson, Lopez Island.

William H. (Harrison) Higgins (1848–1928): born in Illinois; married *Matilda Jane King* (1853–1924) in 1876; homesteader (1881), Beaverton Valley, San Juan Island.

N. P. (Norman Peter) Hodgson (1867–1934): born in Ontario; stepson of William Graham; married *Charlotte Lila "Lillie" Schmaling* (1871–1947) in 1895; storeowner and farmer, Richardson, Lopez Island.

Augustus Hoffmeister (1829–1874): German; post sutler, English Camp, San Juan Island; sheepman, San Juan and Spieden Islands.

Sidney Hudson (1867–1911): married *Janey E. Maloney* (1869–1962) in 1890; farmer, Lopez Island.

Hiram E. (Edson) Hutchinson Jr. (1831–1881): born in Vermont; married *Marion Bone* (1835–1895) ca. 1866; trader and storekeeper, homesteader (1879), and farmer, Lopez Village, Lopez Island.

George P. Jeffers (1902–1971): owner, San Juan Canning Company (pea canning), Friday Harbor, San Juan Island.

Chris Jensen (ca. 1853–?): Danish; farmer, Center Valley, Lopez Island.

Peter Jewell (1826–1876): homesteader (1879) and farmer, San Juan Valley, San Juan Island.

John Kay (?–1907): in 1882 shot his neighbor John Anderson to death over a "breachy cow"; homesteader (1882), Lopez Island.

John Keddy (1833–1907): British; born in Exeter, Ontario; brother of William Keddy; preemptor (1877); sheepman, Cady Mountain, San Juan Island.

William Keddy (1844–1919): British; born in Port Hope, Ontario; brother of John Keddy; sheepman, Lopez Island.

James F. (Franklin) King (1857–1932): born in Yamhill County, Oregon; son of Francis and Sarah King; arrived in the San Juan Islands in 1877; married *Adeline Verrier* (1864–1942) of Westsound, Orcas, in 1880; farmer and fruit grower, Friday Harbor, San Juan Island.

John William King (1859–1953): born in Yamhill County, Oregon; son of Francis and Sarah King; married *Marcia Henri Dightman* (1869–1951) in 1885; farmer, San Juan Island.

Joseph "Joe" T. LaChapelle (1886–1942): Canadian; married *Mary Jane Firth* (1870–1939) in 1907; farmer, American Camp, San Juan Island.

Jim Lawrence (1951-) and Lisa Nash (1959-): farmers, Thirsty Goose Farm (est. 1974), San Juan Island.

Julien (Julian) Lawrence (1837–1905): French Canadian; naturalized in 1873; married *Terice* (1849-1950) in 1878; settler, Shaw Island.

Alfred Lawson (1868–1941): son of Peter Lawson and Fannie Dearden; married *Esther Lucretia Fowle* (1882–1951) in 1905; homesteader (1891) and farmer, West Valley, San Juan Island.

Gilbert "Bert" Joseph Lawson (1915–1981): son of Alfred and Esther; married *Adah Geraldine "Jeri" Halvorsen* (1924–2015); farmer, San Juan Island.

John Lawson (1855–1916): Swedish; married *Jennie Johnson* (1859–1919) in 1891; rhubarb grower, Deer Harbor, Orcas Island.

John R. Lawson (1913–1983): married *Dorothy May Dougherty* (1918–2013) in 1936; farmer, San Juan Island.

Lizzie Lawson (1879–1968): daughter of Peter Lawson and Fanny Dearden; farmer, San Juan Island.

Peter Lawson (1827–1927): Danish; married *Fanny Dearden* (1841–1900); father to Lizzie Lawson; skipper and farmer, San Juan Island.

William W. (Wesley) Lee (1861–1924): from Iowa; married *Cora Belle Brusha* (1864–1934) ca. 1891; rhubarb grower, San Juan Valley, San Juan Island.

Benjamin "Ben" Lichtenberg (1876–1925): born in Pennsylvania; married *Catherine "Kate" Buckley* (1877–1959); established GEM Farm (1898), Lopez Island.

M. R. (Mathias Rudolph) Lundblad (ca. 1819–ca. 1899): Swedish; naturalized in 1876; homesteader (1877) and farmer, Argyle, San Juan Island.

Ira D. (Doen) Lundy (1867–1932): Canadian; arrived on Lopez Island in 1889; married *Mary Ida Stewart* (1868–1938) in 1897; postmaster and berry farmer, Richardson, Lopez Island.

Daniel Madden (1835–1914): Irish; arrived on San Juan Island in 1862; married *Mary Ellen Gorman* (1857–1890) in 1882; homesteader (1877) and farmer, San Juan Valley, San Juan Island.

Patrick Madden (1842–1919): Irish; married *Agnes Catherine Flynn* (1864–1946) in 1881; homesteader (1882) and farmer, San Juan Valley, San Juan Island.

Henry Mathesius (1896–1948): farmer, Crow Valley, Orcas Island.

Dana McBarron (1914–1994): from Montana; married *Frances Henderson* (1914-1996) in 1941; raised Scotch Highland cattle, Cape St. Mary Ranch, Lopez Island.

W. P. McCaffray (1879–1944): founder of the National Fruit Canning Company; establisher of Olga Strawberry Barreling Plant (1936), Olga, Orcas Island.

John S. (Stafford) McMillin (1855–1936): from Indiana; married *Isabella Louella Hiett* (1857–1943) in 1877; established Bellevue Poultry Farm, Roche Harbor, San Juan Island.

George Meyers (1853–?): fruit grower, Orcas, Orcas Island.

William Miller (1828–1901); preemptor (1879) and fruit grower, Crow Valley, Orcas Island.

J. E. (John Elmer) Moore (1872-1930): born in New York; married *Georgia M. Stratton* (1877–1942) in 1899; founder, Orcas Fruit Company (1907), Eastsound, Orcas Island.

Thomas Mulno (1831–1902): from Maine; married *Amanda Clark* (1833–1913); homesteader (1882) and farmer, Mulno Cove, San Juan Island.

Peter Niels Nielsen (1926–2008): Norwegian; farmer, Center Valley, Lopez Island.

Andrew Nordstrom (1850–1928): married *Hannah Johanson* (1855–?); arrived on Orcas Island in 1901; homesteader (1901), farmer, and fruit grower, Crow Valley, Orcas Island.

Charles Victor Peterson (1874–1950): son of P. E. and Christina Ann Peterson; married *Caroline "Carrie" Mathesius* (1883–?); father of Oscar Peterson; farmer, San Juan Island.

Oscar Peterson (1910–1996): son of Charles Victor and Carrie Mathesius Peterson; farmer, San Juan Island.

P. E. (Peter Elmgren) Peterson (1843–1922): Swedish; arrived in the United States ca. 1866; naturalized in 1873; married *Christina Ann McKenzie* (1850–1934) in 1873; father of Charles Victor Peterson; homesteader (1879) and farmer, San Juan Valley, San Juan Island.

Roy P. (Palmer) Prestholt (1906–1987): from Iowa; turkey farmer, Lopez Island.

Henry Quinlan (ca. 1832–1880): Irish; naturalized in 1877; homesteader (1880) and farmer, San Juan Valley, San Juan Island.

M. J. (Martin John) Reddig (1853–1915): from Ohio; married *Sarah Martin Auld* (1857–1922) in 1877; moved to Orcas Island in 1891; builder of farm structures (barns and fruit driers) and orchardist, Eastsound, Orcas Island.

George Stillman Richardson (ca. 1847–1915): born on Mount Desert Island, Maine; married *Ellen Bishop* (1850–1909) in 1871; homesteader (1879) and farmer, Richardson, Lopez Island.

Albion "Ab" K. Ridley (1842–1928): from Maine; married *Elizabeth Graham* (1851–1926) in 1867; arrived on Lopez Island in 1902; farmer and hotelkeeper, Richardson, Lopez Island.

Glen Hubert Rodenberger (1888–1973): married *Elizabeth "Lizzie" Bryan Unbrell* (1897–1987) in 1914; founding member, Orcas Island Berry Growers Association; grower of Marshall strawberries, Olga, Orcas Island.

Christopher Rosler (1840–1906): German; US Army soldier stationed at American Camp 1859–1861; married **Anna Pike** (1846–1909), Tsimshian, in 1861; homesteader (1877) and farmer, south end of San Juan Island.

Francois "Frank" Phillip Rouleau (1880-1936): French Canadian, born in Quebec; arrived around 1904; married **Clara Josephine Marcotte** (1881-1923); established a dairy which delivered raw milk throughout San Juan Island.

L. L. (Linn Legrand) Salisbury (1868-1964): born in Wisconsin; married **Elzada Wilson** (1882–1964) in 1906; dairyman, Friday Harbor, San Juan Island.

Isaac Sandwith (1852–1923): English; arrived on San Juan Island in the 1860s; married **Sarah Harriet Potter** (1855–1933) in 1873; naturalized in 1873; farmer, San Juan Island.

Richard Scurr (1831–1909): English; naturalized in 1870; farmer, White Point, San Juan Island.

Robert (Frederick) Scurr (1833–1913): English; naturalized in 1870; married **Annetta "Nettie" Hill** (1868–1935); homesteader (1889) and farmer, White Point, San Juan Island.

Peter "Whispering Pete" Sery (Seary or Serry) (1839-): Irish, emigrated in 1859; married **Lizzie Davis** (dates unknown) in 1881; homesteader (1899) and farmer, Waldron Island.

Richard J. Shaw (1868–1933): arrived on Shaw Island in the 1890s; married *Janette Ellen Jennie Gordon* (1873–1924) in 1892; farmer, Shaw Island.

D. B. (Daniel Bair) Shull (1851–1926): married **Emmereldes "Emma" Shade** (1857–1941) in 1882; father to Howard and Rebecca Anna Laura; arrived on San Juan Island in 1895 from Port Townsend, Washington; farmer and San Juan County Commissioner, San Juan Island.

Howard Shull (1886–1923): son of Daniel B. and Emma Shull; married *Sarah Estella Newhall* (1889–1919) in 1916 and *Loretta Mary Newhall* (1886–1969) in 1921; farmer, San Juan Island.

Milo Milton Nels Smoots (1874–1953): from Minnesota; married *Sarah E. Middleton* (1879–1972) ca. 1906; temporary secretary, San Juan Dairyman's Association; farmer, San Juan Valley, San Juan Island.

Lawrence Glenn Stanbra (1893–1981): married *Rebecca Anna Laura Shull* (1888–1967), daughter of neighbors Daniel and Emma Shull, in 1920; farmer, San Juan Island.

Albion "Al" Evans Sundstrom (1918–2014): son of John A. Sundstrom and Josephine "Josie" Hazel Madden; married *Winnifred Margaret Kuljis* (1916–1990); married *Deanna M. Spooner* in 1994; farmer, San Juan Valley, San Juan Island.

Clyde Madden Sundstrom (1920–2006): son of John A. Sundstrom and Josephine "Josie" Hazel Madden; married *Ruth Marjorie Guard* (1920–1997), daughter of Roy and Myrtle Guard; farmer, Valley View Farm, San Juan Valley, San Juan Island.

Johan "John" A. (Abdon) Sundstrom (1884–1952): born in Sweden; married *Josephine "Josie" Hazel Madden* (1890–1992) in 1914; father of Al and Clyde; farmer, San Juan Valley, San Juan Island.

W. E. (William Evert) Sutherland (1848–1926): Canadian; moved to Orcas Island in 1883; established Orcas Landing; storekeeper, postmaster, and fruit grower, Orcas, Orcas Island.

C. (Charles) H. Sutton (1841–1913): from England; married *Anna Agnes Lutrick (Lotritz?)* (1850–1933) in 1868; arrived on San Juan Island in 1892; hop farmer, San Juan Valley, San Juan Island.

John Sweeney (1833–1914): Irish; brother of Joseph Sweeney; married *Johannah (Hannah) Antonia? Jane Burkhart* (1845-) in 1865; farmer, San Juan Valley, San Juan Island.

Joseph Sweeney (1841–1920): Irish; brother of John Sweeney; married *Alice Lucinda Boyce* (1860-1943) in 1880; merchant, real estate broker, and farmer, Orcas and San Juan Islands.

Theodore Tharald (1857–1926); Norwegian; brother of Thomas Tharald; arrived from Port Gamble, Washington, in 1883; homesteader (1883) and farmer, Shaw Island.

Thomas Tharald (1854–1923): Norwegian; brother of Theodore Tharald; arrived on Shaw Island in 1885; farmer, Shaw Island.

John Tod Jr. (1845–1889): Scots/Thompson River Indian; son of John Tod, Chief Trader, Hudson's Bay Company; sheepman, Spieden and Henry Islands.

Harry Towell (1862–1936); English: married *Lillian May Stevens* (1868–1950) in 1884; arrived on Lopez Island in 1898; farmer, Center Valley, Lopez Island.

Samuel James Trueworthy (1835–1876): from Maine; married *Jeanine (Jane or Jenny) Sluequitti* (1852–1924) in 1866; sheepman, Orcas Island.

Clarence M. Tucker (1863–1933): born in Indiana; married *Marie Elizabeth Jensen* (1867–1948) in 1900; mill operator and postmaster, Argyle, San Juan Island; later San Juan County Treasurer; fruit raiser, Lopez Island.

James Francis Tulloch (1848–1936): married *Nancy Anne "Annie" Brown* (1856–1941) in 1876; homesteader (1907), farmer, and orchardist, Orcas Island.

C. H. (Clarence Herschel) Van Sant (1868–1947): from Tennessee; married *Francis Mina Robb* (1869–1963) in 1890; nurseryman, Eastsound, Orcas Island.

E. L. (Ernest Louis) Von Gohren (1851–1931): from Tennessee; married *Mary Emma Fry* (1861–1934) in 1878; arrived on Orcas Island in 1879; civil engineer and San Juan County surveyor; orchardist and nurseryman, Eastsound, Orcas Island.

Jessie Murphy Waldrip (1869–?): farmer, Eastsound, Orcas Island.

Edward D. (Dunlop) Warbass (1825–1906): one of the founders of Friday Harbor and the first San Juan County Auditor; farmer, Idlewild, Friday Harbor, San Juan Island.

Bertram "Burt" Weeks (1876–1953): son of Lyman and Irene Weeks and brother of Edson and Oscar; farmer, Lopez Village, Lopez Island.

Edson Weeks (1863–1929): son of Lyman and Irene Weeks and brother of Burt and Oscar; farmer, Lopez Village, Lopez Island.

Lyman Weeks (1820–1900): born in Vermont; married *Irene Melinda Hutchinson* (1840–1926), sister of Hiram Hutchinson, in 1862; father of Burt, Edson, and Oscar; arrived Lopez Island in 1874; farmer, Lopez Village, Lopez Island.

Oscar Weeks (1873–1962): son of Lyman and Irene Weeks and brother of Burt and Edson; married *Henrietta Lee Blake* (1873–1956) in 1896; farmer, Lopez Village, Lopez Island.

John Henry Wilson (1877–1947): married *Lena Boulton* in 1897; farmer and blacksmith, Center Valley, Lopez Island.

J. (John) T. Wright (1854-1936): born in Ontario; farmer, Lopez Island.

Fred Zylstra (1902–1968): Dutch; married *Rena Dorris Hambley* (1900–2005); raised pure-bred, registered polled Herefords, Wooden Shoe Farm, San Juan Valley, San Juan Island.

Index

A

Ackley, A. F., 44–46, 189, 206
Adams, Mike L., 206
Adkins, George H., 77, 113, 206
Anderson, John, 50, 52, 184, 206, 215
apples, 34, 47, 55–58, 60, 73–76, 78, 80, 82-88, 111–112, 127, 135, 151–153, 155, 184, 193
Archambault, J. C., 31, 206–207
Archambault, John, 33, 176, 207
Argyle, 77, 79, 103, 105, 144, 163–166, 177, 179, 194
Arnt, Charles E., 117, 207
Auld, Robert, 155–156, 207

B

Bailer, E. P., 105, 171, 207
Bailer, Gregory, 171, 207
barley, 15, 20, 39, 58, 69–70, 72–73, 101, 118–119, 165, 176, 181
barns, 15, 18, 21, 31, 36–38, 45, 47, 51, 57–60, 68, 70, 80, 90, 92–95, 97, 100, 115, 131, 133-135, 137-149, 152-157, 164, 166–186, 191, 197-198, 199–201
Bartlett, John H., 71, 148, 207, 212
Beaverton Valley, 61, 69, 119, 131, 178–179, 194, 213–215
beef (cattle), 19, 34, 73, 117–119, 124, 126–127, 170, 174, 176, 181
Beigin, Patrick, 31, 105, 170, 207
Belle Vue Sheep Farm, i, iv, 9–15, 17–18, 21, 23–24, 26, 38, 53, 64, 70, 105, 161, 168–169, 182, 188, 193–194, 198, 211–213
Bellevue Poultry Farm, 56, 98–99, 184, 193
Bellingham, v, 3, 33, 82, 88, 92, 107–108, 111–113, 142, 152, 187, 190

Berkshire boar, 19–20
berries, 84–85, 127, 158
Biendl, John, 134, 207
Bjork, August, 34, 207
blacksmith shop, 143
Blake, James, 138–139, 207
Blake, T. J., 137, 139, 207
Blake, Thomas George, 207
bloody murder, 52
Bostian, Peter, 105, 208
Boulton, George Washington, 143, 208
Boyce, Stephen, 166, 171–172, 208, 210
Breedlove, Albert M., 109, 208
Bruns, John E., 108–109, 208
Buchanan, Gordon, 174, 208
Buchanan, James, 59, 208
Bull, Catherine, 26, 31–32, 208
bunkhouse, 175
Burt, John, 47, 208–209
Burt, Joseph, 140, 209
Buxton, C. B., 77, 209

C

Caines, Robert M., 91, 209
camas, i, iv, 7–9, 13, 20, 24–25, 137, 151, 168, 188, 202
cannery, 54, 56–57, 59-60, 84, 102, 104, 118, 145, 162, 214
canning, 60, 85, 104, 118, 157, 162, 215
Cantine, C. E., 74–75, 79, 82, 209
Cantine, Edward "Ed" I., 74–75, 79, 209
Capron, Victor, 182, 209
cattle, 4, 14–15, 19, 21, 25, 31–32, 34, 41, 46, 48–49, 55–56, 62–63, 73, 86, 92, 94, 104, 117–119, 124, 126, 135, 139, 142, 146, 148, 154–156, 161, 164, 166–168, 173–174, 176–178, 180, 183, 200, 209

Center Valley, 47, 69, 137, 142, 208–209, 211, 215
Chalmers, Albert "Bert" Earnest Murray, 154, 209
Chalmers, Estyn M. (Murray), 209
Charnley, Brent, 139, 209
cherries, 47, 55–56, 58, 74–76, 84, 135, 151, 153
chicken coop, 98, 134–135, 138, 143, 172, 180
chicken house, 33, 40, 45, 47, 141–142, 155, 166, 175
chickens, 31, 34, 40, 43, 46, 56–57, 62, 98–100, 112, 127, 176, 181–182, 184
Christensen, Hans Lee, 37, 210
Clark, W. O., 43, 77, 193, 210
climate, iv, 1, 4–6, 15, 41, 67, 104–105, 129, 187
Coast Salish, i, iii–iv, 7–9, 24, 31, 33, 37, 42, 49, 111, 137, 139, 151, 161, 187–188, 202–203
Coffelt, Vern, 125–127, 194, 210
Community Supported Agriculture, i, 128
conservation district, 115, 120, 192
county agent, 101, 113, 165, 191
Cousins, James and John, 47, 88, 145–146, 210
cows, 10, 19, 47, 52, 55–56, 58, 73, 86, 88, 90–97, 113, 140, 144, 147, 149, 154, 157, 167, 172, 175–177, 180-181, 184,197, 200, 206, 215
creamery, 59–60, 86–88, 90, 92, 117, 146, 162, 185, 197
Crook, James "Jim", 184, 210
Crook, William, 183, 210
Crow Valley, 3, 33, 47, 69, 77, 80, 91, 104, 126, 131, 151, 153, 155, 207, 209–210, 212, 214
Culver, George N., 77, 210
Cutlar, Lyman, i, 20, 23–25, 105, 167, 210

D

dairy, v, 19, 21, 34, 38, 40, 55, 67, 73, 86–95, 99, 104, 113, 116–119, 127–128, 134–135, 139–140, 142–144, 146–149, 153–154, 156, 164, 174–178, 180, 184, 197, 200–201, 210

Davis, Tommy, 59, 61, 111, 119, 177, 183, 191, 210
De Laval, 91, 97
Deer Harbor, 3, 33, 108–109, 151, 194, 214, 216
Douglas, Alfred, 42, 101, 210
Douglas, James, 10, 13–16, 210
Douglas, Robert, 210–211
Doyle, Frank, 185
Doyle, John P. (Patrick), 175, 211
draft horse, 58–59, 61, 116, 194
drainage ditches, 131
Driggs, G. B., 77–78, 163, 194, 211

E

Eastsound, vi, 3, 33, 47, 55, 60, 73–74, 77, 80, 82, 87, 112, 125, 131, 151, 156, 191, 194, 206, 208, 214
eggs, 33, 38, 56, 58, 79, 82, 98, 100, 111, 127, 181–182, 184
Ender, Joseph, 147, 211
Entermann, Andre, 127, 142, 211
Erb, Henry, 148–149, 211
Extension Agents, vi, 66, 196, 205

F

farmhouse, 47, 147, 153–154, 164, 166–167, 171, 175–178, 181
fences, 15, 18, 27, 45, 48–54, 80–81, 128, 131–132, 134, 182, 195
Firth, Robert, 35, 53–54, 63, 71, 169, 188, 194–195, 211, 214, 216
Fleming, Thomas, 43, 49, 53, 105, 189, 211
Fluid Milk Law, 117
Fowler, Elihu B. & Oscar William, 40, 100, 120, 155–156, 212
frame construction, 36, 40, 80, 147, 154, 164, 175, 177, 197, 200
Frazer, Robert & William, 26, 167, 212

Friday Harbor, 11, 33, 47, 55, 60, 64–65, 72, 77, 79–80, 85–86, 88, 92, 97, 100, 102–105, 108, 110–113, 115–118, 120, 125–127, 134–135, 143, 146, 161–163, 165, 179–180, 184–185, 190–191, 194, 208, 211, 213–215
Friday, Joe & Peter, 172, 212
fruit, ii, v, 21, 27, 34, 41, 43, 55–58, 60–62, 73–78, 80–85, 91, 104, 107, 112–114, 127, 135, 140, 144, 151–158, 163–165, 169, 181, 184, 189, 197, 201, 208, 210–212, 214–215
fruit barns, 80, 201
fruit dryer, 77, 152, 197

G

Gallanger, Joseph, 148
Gallanger, William Walter, 148, 212
Gann, Ernest, 174, 212
GEM Farm, 55–57, 98, 134–135, 193, 216
Gibbs, George William, 82, 106–107, 113, 152–153, 188, 212
Gilmer, Art, 179–180, 212
ginseng, v, 104, 108–109, 126, 208
goats, 8, 63–64, 127, 142–143, 203
Goodrow, Barney, 84, 144, 212
Gorman, Joseph J. & Patrick, 213
Gorman, Peter, 31, 174, 179, 213
Graham, Thomas & William, 47, 145–146, 213, 215
grains, v, 14–15, 18, 20, 25, 27, 32, 36–37, 39, 42, 47, 58, 61–62, 69–72, 86, 100, 103–104, 119, 141–143, 162, 164–165, 169–171, 173, 175, 183, 197, 200
granary, 18, 21, 33, 39, 70, 100, 131, 138, 141–143, 145, 148, 164, 166–175, 177–181, 197
Grange, vi, 114–115, 141, 144, 163, 208
Griffin, Charles John, i, 5, 9–11, 13, 16–21, 23–24, 26–27, 44, 53, 62, 70, 105, 165, 168, 170, 194, 213
Griswold, George Oscar, 213
Groll, Joe, 97, 186, 213

Guard, Frank, 213
Guard, Harold, 178, 213
Guard, Leroy "Roy" Paul, 176, 214
Guard, Paul, 69, 176, 178, 194, 214

H

Hannah, Harriet Delila "Lila", 214
Hannah, James, 54, 63, 214
Harrison, Ben, 113, 178, 214–215
Harvestore, 97, 148–149
hay, v, 6, 14–15, 20, 24, 32, 36–38, 41, 45, 47, 58, 62, 67–69, 94, 97, 104, 119, 127, 132, 134, 138–144, 146–147, 149, 153–154, 157, 165–166, 171–174, 176–181, 197–200
hay rail-and-trolley system, 68, 147, 178, 197
Hayton, William, 149, 214
Heidenrich, George, 100, 214
Henry, John"Pea", 102, 104, 143
Hershberger, George L., 107, 109, 194, 214
Higgins, Owen James, 146, 214
Higgins, William H., 215
Hodgson, N. P., 47, 145, 215
Hoffmeister, Augustus, 62–63, 65, 183, 215
hogs, 25, 31, 34, 40–41, 45–46, 48–49, 55, 92, 148, 180
homestead, v–vi, 23, 26, 29, 31–34, 37–38, 40–41, 44–48, 51–53, 73, 86, 132, 137, 139, 145, 155–156, 163, 167, 169–171, 173, 182–183, 189, 206
hops, 77, 104–106, 170
Hudson's Bay Company frame, 22, 198
Hudson, Sidney, 148, 215
Hutchinson, Hiram, 63, 139–140, 215

I

Italian prune plums, 73, 156

J

Jeffers, George P., 85, 118, 162, 215
Jensen, Chris, 141, 215
Jewell, Peter, 33, 175–176, 215

K

Kay, John, 52, 206, 215
Keddy, John & William, 181–182, 215
King, James F. (Franklin), 163–165, 215
King, John, 185, 216
kitchen garden, 18, 21, 40–41, 181

L

LaChapelle, Joe, 105, 216
Lawrence, Jim, 109, 216
Lawrence, Julien, 37, 216
Lawson, Alfred, 180, 182–183, 216
Lawson, Gilbert "Bert" Joseph, 180, 216
Lawson, John, 108, 216
Lawson, John R., 166–167, 216
Lawson, Peter, 169, 182, 216
Lee, William W., 108, 216
Lichtenberg, Ben, 55–56, 98, 135, 216
loafing shed, 147, 173
Lopez Island, ii, 1, 3–4, 29–30, 33, 43, 47–48, 52, 54–57, 59–62, 69–71, 74, 83–84, 86–88, 91, 100, 103, 110–111, 113, 115–117, 124–125, 127–128, 130–132, 134–149, 168, 182
Lopez Village, 33, 74, 125, 127, 137, 139–140, 143, 209, 215
Lundblad, M. R., 105, 217
Lundy, Ira D., 84, 145, 217

M

machine shed, 47, 138, 141–142, 146–147, 153–154, 166–167, 169, 173
Madden, Daniel & Patrick, 33, 217
marijuana, 128
Mathesius, Henry, 153, 166, 217
McBarron, Dana, 117, 217
McCaffray, W. P., 85, 217
McMillin, John S., 57, 98–99, 184, 217
Metcalf, Elizabeth, 211
Meyers George, 82, 113, 152–153, 217
milk house, 47, 90–93, 131, 134–135, 139, 141–143, 146–149, 153–154, 164, 166–167, 174–178, 181, 198
milking parlor, 143, 172–173, 179, 185
Miller, William, 110, 153, 217
mink, 129
Moore, J. E., 56–57, 59–60, 217
Morrill Tariff, 64
mow, 37, 67–69, 134, 138, 140, 166, 174, 179, 197–198
Mulno, Thomas, 34, 43–46, 206, 217

N

Nash, Lisa, 216
Nielsen, Peter Niels, 147, 218
Nilan, Maggie, 139, 209
Nordstrom, Andrew, 153–154, 218

O

oats, 15–16, 18, 20, 31, 39, 58, 69–73, 88, 118–119, 165, 168–169, 171, 176, 178, 194
Olga, 4, 6, 47, 82, 84–85, 113, 151, 157–158
Olga Strawberry Barreling Plant, 84, 157

Olympia, v, 1, 50, 76
open range, 24, 47–50
Open Space, 121–122
Orcas Fruit Ranch and Cannery, 57
Orcas Island, ii, vi, 1, 3–5, 29–30, 33, 43, 47, 54–55, 57, 60, 63, 65, 69, 73–74, 77–82, 85–86, 91, 93, 101, 104–110, 112–117, 120, 124–127, 130–132, 134–135, 150–159, 163, 177, 206–210, 212, 214–216
organic, 124–125, 139
outhouse, 35, 134–135, 175, 179
oxen, 18–20, 35, 58, 73

P

Pacific Northwest Guernsey Breeders Association, 88–89, 194
peaches, 74, 76
pears, 34, 47, 55–58, 73–74, 76, 81–84, 135, 151–153
peas, v, 15, 18, 20, 33, 46, 69–72, 102–104, 112, 118, 142, 162, 165, 168, 170, 174, 177, 195, 214–215
pests, 42, 79, 203
Peterson, P. E., 166, 218
pigs, 15, 19–20, 23, 25, 181–182
plums, 47, 55–56, 58, 70, 73–74, 76, 78, 135, 153, 156, 163
pole construction, 198
ponds, 27, 119–120, 128
Port Stanley, 148, 212
Port Townsend, 24, 44, 70, 82, 105, 165, 179
Post, Dodie, 174, 212
potatoes, i, v, 7, 15, 18, 21, 23–25, 29, 33, 38, 41, 46, 48, 58, 104–105, 167–169, 171, 200
poultry, v, 34, 40, 53, 55–58, 62, 98–100, 112–113, 135, 137–139, 156, 165, 173, 176, 184, 193, 201
predators, 17, 20, 23, 42–43
Prestholt, Roy P., 137, 218
prune dryer, 77, 153, 155–156, 193

prune plums, 70, 73, 153, 156
prunes, 57–58, 74, 76–79, 82–83, 151, 163, 165
Public Land System, 28, 199

Q

Quinlan, Henry, 31, 174, 218

R

rabbits, 101, 104, 109–110
Reddig, M. J., 77, 153, 218
rhubarb, ii, v, 104, 107–109, 127, 194, 214, 216
Richardson, 47, 59, 84, 137, 144–146, 191, 213–215
Richardson, George Stillman, 218
Ridley, Ab, 47, 145–146, 218
Rodenberger, Glen Hubert, 85, 218
root cellar, 38, 47, 134–135, 142, 146–148, 164, 167, 178, 200–201
root house, 18, 33–34, 38, 45, 157, 167, 184, 200
Rosler, Christopher, 31, 167, 219

S

Salisbury, L. L., 88, 219
San Juan County Fairgrounds, 112, 165
San Juan Island, iv, 1, 3–4, 10–11, 13–16, 23, 26–33, 43–44, 54–55, 57, 60–66, 69–70, 77, 98, 100–105, 107–129, 144, 158, 160–186, 206–222
San Juan Valley, 3–4, 14, 26, 30–33, 42, 58, 61, 85, 100, 103–105, 108, 110, 117–121, 124, 131, 161–162, 166, 169–170, 172–174, 176–177, 181, 191, 194–195, 206–208, 210–216
Sandwith, Isaac, 63, 171–172, 183, 219
Scurr, Richard & Robert, 184, 219

Seattle, v, 77, 82–83, 86, 88, 100, 105, 108, 114, 116, 156, 187, 190
seed crops, 104, 110
Sery, "Whispering Pete", 80, 219
Shaw Island, 1, 34, 37–38, 40–41, 46–47, 53, 55, 60, 77, 87, 100, 105, 109, 114, 124, 128, 130, 134–135, 207–208, 210, 212–213, 216
Shaw, Richard J., 134, 219
sheep, i–ii, iv–v, 8–21, 23–26, 31–34, 38, 41, 43, 46–50, 53–55, 62–67, 70, 73, 104–105, 112, 118, 124, 126–127, 155, 161–162, 164, 167–170, 176, 181–183, 188, 193–194, 198, 209–213
Shull, D. B., 179, 219
Shull, Howard, 92, 180, 219
silo, 96–97, 131, 134, 138–139, 148–149, 157, 176–177, 179–180, 200
Smith-Lever Act, 110, 112
smokehouse, 40, 47, 134–135, 142, 146–147, 155, 174–175, 200
Smoots, Milo, 219
soil, iv, 1–3, 6–7, 14, 18, 20, 23, 27, 35, 38, 41, 45, 58, 61, 69, 84, 86, 91, 107–108, 110, 115, 118, 120–121, 124, 128–129, 131–132, 137, 147, 151, 165, 170, 178, 187, 191, 195
stable, 19, 21, 34, 40, 110, 139, 185
stalls, 59–60, 68–69, 90, 94–95, 97, 140–141, 143, 146–147, 149, 157, 166, 174, 177, 180, 183, 200
Stanbra, Lawrence Glenn, 179, 220
stanchions, 90, 94–95, 97, 134, 140, 142–144, 146–149, 153–154, 157, 166–167, 172, 174–181, 185, 200
strawberries, 58, 74, 84–85, 157–158, 163
Stuart Island, 1, 9, 124
Sundstrom, Albion "Al" Evans, 220
Sundstrom, Clyde Madden, 176, 220
Sundstrom, Johan "John" A. (Abdon), 220
Sutherland, W. E., 151, 220
Sutton, C. H., 106, 220
Sweeney, John, 33, 176–177, 220
Sweeney, Joseph, 32–33, 220

T

Tacoma, v, 57, 82, 184
Tharald, Theodore & Thomas, 46, 53, 221
thresher (threshing machine), 59, 71–72, 164, 207–208
timber-frame construction, 200
Timothy grass, 14, 24, 41, 203
tobacco, 16–17, 104–105
Tod, John Jr., 63, 183, 221
Towell, Harry, 127, 142, 221
Trueworthy, Samuel James, 62–63, 221
Tucker, Clarence M., 31, 77, 144, 186, 221
tulips, v, 104, 106–107, 152
Tulloch, James Francis, 50, 65–66, 74–75, 79, 82–83, 157, 189, 221
turkeys, 55, 57, 62, 98–101, 112, 135, 137, 176, 184, 193, 214
turnips, 15, 18, 20, 33, 38, 41, 46, 48, 200

V

Van Sant, C. H., 221
ventilators, 77, 80, 94, 153–154, 157, 181, 185, 201
Victoria, BC, v, 5, 10, 13, 19–21, 26, 28, 31, 33, 42, 62–63, 65, 76, 82, 111–112, 171, 190, 211
Von Gohren, E. L., 78, 83, 189, 221

W

Waldrip, Jessie Murphy, 81, 115, 156, 193, 221
Warbass, Edward, 48, 221
water towers, 114–115, 131, 140, 157, 179–180, 193, 201
weeds, 20, 42, 105, 118, 165, 203–204
Weeks, Bertram "Burt", 140, 221
Weeks, Edson, Lyman, & Oscar, 221–222
well house, 138–139
West Sound, vi, 60, 77, 82, 93, 113–114, 153, 163

wheat, 15, 20, 48, 58, 69–72, 119, 165, 171, 181
Wilson, John Henry, 143, 222
wineries, 127
woolly dogs, i, iv, 8–9, 49, 151, 188, 193
worm fences, 51–52
Wright, J. T., 72, 222

Z

Zylstra, Fred, 117, 120–121, 173–174, 222

www.ingramcontent.com/pod-product-compliance
Lightning Source LLC
Chambersburg PA
CBHW071957070526
44583CB00015B/1234